OXFORD
UNIVERSITY PRESS

PROGRESS

ESSENTIAL

PHYSICS STAGE 9

FOR CAMBRIDGE SECONDARY 1

Darren Forbes

Viv Newman | Editor: Lawrie Ryan

OXFORD
UNIVERSITY PRESS

Great Clarendon Street, Oxford, OX2 6DP, United Kingdom

Oxford University Press is a department of the University of Oxford.
It furthers the University's objective of excellence in research, scholarship,
and education by publishing worldwide. Oxford is a registered trade mark of
Oxford University Press in the UK and in certain other countries

First published by Nelson Thornes Ltd in 2013
This edition published by Oxford University Press in 2015

British Library Cataloguing in Publication Data
Data available

978-0-19-839992-6

10 9 8 7 6 5 4

Printed in Great Britain by CPI Group (UK) Ltd., Croydon CR0 4YY

Acknowledgements

Cover photographs: Getty Images/Eric Meola
Illustrations: Tech-Set Ltd
Page make-up: Tech-Set Ltd, Gateshead

Although we have made every effort to trace and contact all
copyright holders before publication this has not been possible in all
cases. If notified, the publisher will rectify any errors or omissions at
the earliest opportunity.

Links to third party websites are provided by Oxford in good faith
and for information only. Oxford disclaims any responsibility for
the materials contained in any third party website referenced in
this work.

Contents

Access your support website:
www.oxfordsecondary.com/9780198399926

Photo acknowledgements

Alamy: ClassicStock, p3 (left); Leslie Garland Picture Library, p6; Andrew Kitching, p16. **Fotolia:** p59; p71. **Getty Images:** Gamma-Rapho via Getty Images, p69. **iStockphoto:** p2 (top and bottom left); p3 (right); p7; p8; p9; p20; p22; p23; p25; p30; p39;p41; p43; p44; p51; p54; p55; p57; p61; p64; p65; p66; p68. **Press Association:** ABACA/ABACA, p28 (left). **Rex Features:** KeystoneUSA-ZUMA, p28 (right). **Science Photo Library:** p2 (bottom right); Martyn F. Chillmaid, p14; RIA Novosti, p63; p70.

Welcome to *Science for Cambridge Secondary 1!*
This Student book covers Stage 9 of the Physics curriculum and will help you to prepare for your Progression test and later, your Cambridge IGCSE® Sciences.

Using this book

This book covers one of the main disciplines of science, **Physics**, though you will find overlap between the subject areas. Each chapter starts with Science *in context!* pages. These pages put the chapter into a real-world or historical context, and provide a thought-provoking introduction to the topics. You do not need to learn or memorise the information and facts on these pages; they are given for your interest only. Key points summarise the main content of the chapter.

The chapters are divided into topics, each one on a double-page spread. Each topic starts with a list of learning outcomes. These tell you what you should be able to do by the end of that topic.

Learning outcomes

Key terms are highlighted in **bold type** within the text and definitions are given in the glossary at the end of the book. Each topic has a list of the key terms you should understand and remember.

Key terms

Summary questions at the end of each topic allow you to assess your comprehension before you move on to the next topic.

Summary questions

Expert tips are used throughout the book to help you avoid any common errors and misconceptions.

Expert tips

Practical activities are suggested throughout the book, and will help you to plan investigations, record your results, draw conclusions, use secondary sources and evaluate the data collected.

Practical activity

IGCSE Links prompt you to think forward to what you will learn about at IGCSE level, helping you to prepare for this transition and showing you the importance of studying Stage 9 material in preparation for Cambridge IGCSE.

🔗 *IGCSE Link...*

At the end of each chapter there is a double page of examination-style questions for you to practise your examination technique and evaluate your learning so far.

Answers to Summary questions and End of chapter questions are supplied on a separate Teacher's CD.

Student's website

The website included with this book gives you additional learning and revision resources in the form of interactive exercises, to support you through Stage 9.
www.oxfordsecondary.com/9780198399926

Science *in context!*

Putting the world in your hand

Imagine a day without electricity, a day where the hundreds of electrical devices that surround you don't work. If you'd been born 100 years earlier your life may have been very different.

Shocking starts

Some electrical effects were discovered thousands of years ago by the ancient Greeks. The effects were closely linked to magnetism. Nobody could really think of a useful application for the effects, but hundreds of scientists continued to investigate them as the centuries passed. Progress was very slow as there was no easy way of producing an electric current.

In 1800, Alessandro Volta invented the first battery; the voltaic pile. With this he could produce a steady electric current simply in a laboratory. This breakthrough allowed other scientists to investigate electricity in far more controlled conditions. The idea of electrical circuits progressed, although there still weren't any useful applications.

Useful circuits

Over the next 100 years, knowledge of electricity developed rapidly. The heating effect of a current was noticed, as was the fact that thin wire would glow brightly before melting. Only two years after the development of a battery, Humphrey Davy demonstrated the concept of an electric light based on a platinum wire. It did not operate for long but the idea of electrical lighting was born. Rival groups tested different materials to try to manufacture a glowing wire that would not burn out. Eighty years later Joseph Swan and Thomas Edison developed working lamps using carbon filaments inside glass evacuated bulbs. The lamps were expensive but only a few years later they were replaced by far superior tungsten filaments. At the start of the twentieth century, electrical lighting began to become common.

There are many arguments about who invented the first light bulb

Electricity to everybody

Providing electricity to homes became an important issue. At first only the very rich could afford their own generators. Thomas Edison and Nikola Tesla developed different systems for transmitting mains electricity. Fierce competition broke out between the two rivals. Edison was so determined to prove that Tesla's system was dangerous he publicly electrocuted animals including an elephant! In the end Tesla's technology was far superior at transmitting electricity over long distances and so it was adopted. The mains supply in every country now relies on the ideas developed by Tesla even if there are slight differences between the supplies.

Edison and Tesla were fierce competitors

The availability of electricity everywhere allowed electric motors, electrical heating and electrical lighting to spread. A second industrial revolution began with steam power being rapidly replaced by electrical energy.

Computing

Increased understanding of electrical circuits led to the development of electronic computers. At first these computers filled whole rooms and could only carry out simple programs. The development of miniature components and transistors allowed the computer to become smaller and faster year after year. Computers are now millions of times faster and can store billion of times more information than the earliest machines.

A modern smartphone is far more powerful than this giant computer from the 1960s

Smart everything

Computing technology is now used almost everywhere. Televisions and washing machines have small processors that help them to carry out their functions. Smart mobile phones have much more computing power than the desktop computers from ten years ago. The internet allows billions of computing devices to share information almost instantly so we can find out almost anything from almost anywhere.

The development of computing technology and the long history of our understanding of electricity shows no sign of slowing down. Imagine how powerful computers will be in twenty years' time!

In this chapter you will learn about the idea of electric charge and electric currents. You will build circuits, measure their properties and explore a wide range of electrical devices.

Key points

- Materials can become electrically charged when frictional forces transfer electrons.
- A flow of electrical charge is called a current.
- In a circuit the current is caused by a flow of electrons that carry negative charge.
- Symbols are used to represent components in a circuit.
- Series circuits have only one route for the current to take.
- The greater the resistance in a circuit is, the more the current is reduced.
- Parallel circuits contain more than one route for the current to take.
- Potential differences are produced by chemical reactions in cells.
- Sensors measure environmental changes.
- Digital sensors convert measurements into digital information for processing.
- Direct current travels in one direction and is produced by cells and batteries.
- Alternating current is produced by generators and is provided by mains electricity.
- Mains electricity operates at high potential differences.

1.1 Static electricity

Learning outcomes

After this topic you should be able to:
- describe how materials can become charged with static electricity
- describe observations of the effects that charged objects have on each other
- interpret results using scientific knowledge and understanding.

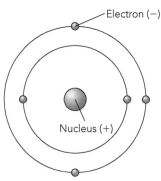

The structure of an atom

Charges and forces

Atoms are composed of three types of particles; protons, neutrons and electrons. Protons and electrons have a property called electrical charge.

- Protons are **positively charged**.
- Electrons are **negatively charged**.

In most materials there is the same number of electrons as protons. Therefore the materials are uncharged overall.

Charged particles produce a force on each other in a similar way to magnets.

- Opposite charges attract each other. (Positive charges attract negative charges.)
- Similar charges repel each other. (Positive charges repel positive charges.)

Separating charges

When some materials are rubbed together, the frictional forces can cause electrons to be transferred from one material to another. Protons cannot move from place to place easily.

- When extra electrons enter an object, the object becomes negatively charged.
- When electrons leave an object, the object becomes positively charged.

Charged objects attract or repel each other according to the rules above.

Practical activity   Charging rods

You have been provided with rods made from different materials, some cloths and a watch glass. Place the watch glass upside down onto the desk.

Charge up one of the rods by rubbing it with a cloth.

Place the charged rod onto the upturned watch glass carefully so that it is balanced.

Charge up a rod of the same material as the first by rubbing it with the cloth.

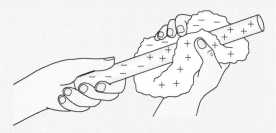

Charges can be separated by frictional forces

Place the second rod near the end of the first without allowing them to touch.

Repeat the experiment with different combinations of rods.

- Describe what happens to the different combinations of charged rods when they are placed near each other.

Induced charge

When a positively charged balloon is placed near an uncharged wall, the balloon will become attracted to the wall. The electrons in the wall are attracted to the positively charged balloon. Some of the electrons in the wall will move towards the surface of the wall causing the surface to become negatively charged.

We say that the wall has a negative charge '**induced**' on its surface. The positively charged balloon is attracted to the negatively charged surface and so it sticks to the wall.

Practical activity   Investigating induced charges

Cut out some very small pieces of paper and place them on a desk. Charge up a plastic rod using a cloth and place the cloth near the paper pieces.
- Describe what happens to the pieces of paper.
- Explain why the charged rod affects the uncharged pieces of paper.

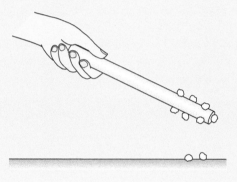

Expert tips

It is the movement of electrons from place to place that causes objects to become charged.

Key terms

- **positive charge**
- **negative charge**

Summary questions

1 Copy the following sentences selecting the correct options.

When a polythene rod is rubbed with a cloth **protons/electrons** move from the cloth to the rod. The rod becomes **positively/negatively** charged and the cloth becomes **positively/negatively** charged.

2 A student walks along a nylon carpet and then reaches out to a door handle. A tiny spark is produced and the student feels an electric shock. Explain why this happens.

3 Dust particles are uncharged but they often stick to electrical equipment such as television screens. Why does this happen?

When large charges build-up, a spark can be produced. A spark is the movement of charge from place to place; we call the movement of charge a **current**.

The Van de Graaff generator

Rubbing a rod with a cloth can only separate a small amount of charge. To produce a larger build-up of charge a Van de Graaff generator can be used.

- A large, hollow, metal dome is insulated from the ground by a plastic tower. The dome needs to be large to allow a large amount of charge to become stored on it.
- Inside the tower a rubber belt is moved by an electric motor.
- The rubber belt rubs against a blade connected to the dome and picks up electrons. The rubber is an insulating material so the electrons are trapped on the belt.
- The electrons are carried by the belt to a brush on the base of the Van de Graaff generator. There the electrons are transferred to the ground.
- As the rubber belt rotates, more electrons are removed from the dome. This builds up a large amount of positive charge on the surface of the dome.

Practical activity   Demonstrating a Van de Graaff generator

Your teacher will demonstrate a Van de Graaff generator (VDG). Listen to the explanations that accompany the demonstration and then answer the questions.

- Why does hair separate when it becomes charged by the VDG?
- What causes the sparks produced by the VDG?

⚠ **The sparks produced by a VDG can be a danger to people with pacemakers or heart conditions.**

A Van de Graaff generator

Lightning

Clouds can become highly charged by frictional forces acting inside them. Droplets of water or ice crystals brush against each other and charges are separated. This process can separate massive amounts of charge. Eventually the charge separation is so large that a lightning bolt is produced.

Most lightning bolts are a flow of electrons directed towards the surface of the Earth. A very large electric current is produced. This current heats the air around it rapidly, causing a bright white glow. The air is heated so quickly that a loud sound shockwave is produced. This shockwave is thunder.

Protection from lightning

Lightning bolts can cause damage to buildings as the electric charge passes through them. To prevent damage, tall buildings often have lightning conductors. Lightning conductors are thick metal cables or bars that allow the electric charge from lightning to pass through safely if a building is struck.

Lightning regularly strikes tall buildings but it can strike anything in its path to Earth – even ships

Key terms

- **current**

Summary questions

1 When you take off nylon clothing there are often tiny sparks. What causes these sparks?

2 Sometimes, when you brush your hair, the hair separates and sticks up. Explain this effect.

3 Why shouldn't you stand beneath a tree in a lightning storm?

The build-up of static charge can be very dangerous or damaging. Static discharges can damage electrical circuits in computers and even small sparks can lead to explosions in some situations.

This electronic engineer is connected to the Earth through a wristband as he repairs a computer. Any static build-up in his body will discharge through the lead instead of damaging the delicate circuitry

Refuelling explosions

Large electrical charges can build up on vehicles when they move. Air resistance or other frictional forces rub against the surface and transfer electrons.

Car tyres can transfer charge to the car's metal body. If enough charge builds up on the car then a spark can be produced. When filling up the car with petrol a small spark can ignite petrol vapour and an explosion can be caused. Modern car tyres have been designed to help reduce the dangerous build-up of charges.

When aeroplanes are refuelled, large amounts of fuel are transferred quickly. The moving fuel could cause a build-up of charge in the aircraft's fuel tanks and a spark could cause an explosion. To prevent this build-up of charge the aircraft and refuelling tanker are connected together by a conducting cable to allow the charge to safely dissipate.

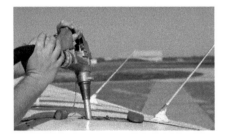

Aircraft can build up static charge as they fly. These thin spikes on the wing are designed to help reduce the build-up of static electricity

Practical activity   Demonstration of sparks causing fire (Bunsen or ethanol)

The sparks generated by static electricity can cause fires and explosions. Watch the demonstration of a Van de Graaff generator being used to light a Bunsen burner or to set fire to some ethanol spilled onto a heat resistant mat.

 This activity must only be performed by a teacher.

Photocopiers and laser printers

Static electricity can also be useful. The effects of forces produced by static electricity are used to improve painting surfaces, clean the air and in photocopiers.

Photocopiers use careful control of static charge to produce images on paper.

- A rotating drum is negatively charged by a brush.
- A bright light is shone onto an image on a sheet of paper.
- The bright light is reflected onto the charged drum.

- The negative charge can escape from the drum in areas that receive the reflected light. The areas that do not receive light remain negatively charged. This means that the charges on the drum are arranged in the same pattern as the image on the original sheet of paper.
- A positively charged 'toner' powder is applied to the drum. This powder is attracted to the negatively charged areas but not to the uncharged areas.
- A negatively charged piece of paper is pressed against the rotating drum. The toner remaining on the drum is attracted to the paper.
- The paper is heated and the toner melts onto it creating a permanent copy of the original image.

A bright light shines onto the original and is reflected onto the rotating drum

Laser printers operate in a similar way. Instead of using reflected light to create the image on the drum, a laser beam draws a pattern so that any image can be created.

The light sensitive drum of a photocopier being repaired

Practical activity Electrostatic research

There are many other uses of electrostatic charges and some additional dangers. Research at least one more application where electrostatic charges are used in a beneficial way and one where the charges present a hazard.

- Present your findings to the class.

Summary questions

1 Why must an electrical engineers discharge themselves before working on electronic equipment?

2 Toner powder is made from very small particles of black plastic. The powder can be very difficult to clean up when spilled. Design a way of cleaning up toner powder using your knowledge of static electricity.

Currents

Electrical charges move around circuits in a current. In an electrical circuit we use a current to transfer energy to electrical devices. In a simple circuit the current is a flow of electrons through metal wires and components. The current is not used up as it travels around the circuit; all of the electrons complete the whole journey releasing energy as they travel.

- The size of a current is the rate of flow of charge and this is measured in a unit called the **ampere** (A).

Potential differences

The current in a circuit is caused by a **potential difference** (p.d.) making the electric charges move. Potential differences can be produced by electrical power supplies and by electrical cells. The size of a p.d. is measured in a unit called the **volt** (V).

Complete circuits

Power supplies and cells have two connections; a positive and a negative terminal. The terminals are indicated with a '+' symbol for the positive terminal and a '−' symbol for the negative terminal. A complete circuit from a positive terminal to a negative terminal is needed for a current to transfer energy. Any gaps in the circuit will prevent the circuit from operating.

Practical activity   Building a complete electrical circuit

Use a battery pack (or power supply), some leads, a switch and a lamp to build a simple electrical circuit. Pressing the switch makes a complete circuit and the lamp should light up if you have connected everything correctly.

- Describe the energy transfer happening in the circuit.
- Examine the switch. Explain the way a switch allows or prevents the current from lighting the lamp.

 Do not use large currents in circuits.

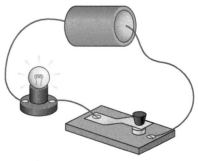

A simple circuit

Conductors and insulators

Electrical charge can pass easily through some materials but not through others.

- It is easy to produce a current in metals so metals are good conductors of electricity. The free electrons within the metal carry charge easily.
- Plastics do not allow a current and so are classed as electrical insulators. Plastics have no free electrons to carry charge.

Practical activity   Testing materials

Use the circuit you built in the previous practical to find out which materials are insulators and which are good conductors of electricity.

Replace the switch in the circuit with the test material.

If the lamp lights up then the material is a conductor.

- Do all of the metal samples conduct?
- Do any of the non-metal samples conduct?

 Do not attempt to use large currents.

Place the material in the circuit to find out if it can conduct electricity

Expert tips

Always build electrical circuits in a logical order. Start at the positive terminal of the battery or power supply and add components one at a time until you reach the negative terminal.

Key terms

- **ampere**
- **potential difference**
- **volt**

Summary questions

1 Copy and complete:

The size of an electric current is measured in _____ (_).

The size of a potential difference is measured in _____(_).

2 Which particles carry charge in electrical circuits? Which type of charge do the particles carry?

3 a) Why are the pins on an electrical mains plug made from metal?

b) Why is the case of the plug made from plastic?

After this topic you should be able to:

- use circuit symbols and diagrams to represent simple circuits
- construct simple circuits from circuit diagrams.

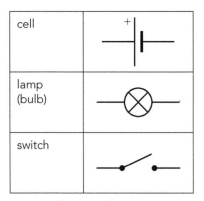

cell	+
lamp (bulb)	
switch	

Some simple electrical components

Electrical engineers do not draw circuits as realistic pictures. This type of drawing would be very time consuming and confusing. Instead, engineers use a standard set of symbols and drawing techniques to produce a **circuit diagram**.

Circuit diagrams are more useful than simple drawings because:

- the same symbols are used around the world allowing all engineers to understand them
- a circuit diagram shows the important connections between components very clearly.

Using symbols to draw circuits

The circuit symbols for a **cell**, a lamp and a switch are shown on the left. The leads or wires that connect components together are drawn as straight lines that meet or cross at right angles. Where two leads join together a dot can be used to show the join clearly.

To draw a circuit clearly follow these simple rules:

- Use a ruler to draw all of the connecting leads and wires and any other straight lines.
- Start by drawing the battery.
- Work around the circuit in a logical order. Start from the positive terminal of the battery.
- Draw one component at a time working towards the negative terminal of the battery.
- If the circuit has any branches then complete one branch before moving on to the next.
- There should be no gaps in the circuit.

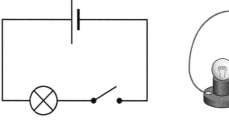 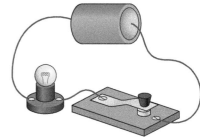

A circuit diagram represents the components in an electrical circuit

Series and parallel circuits

In some circuits there is only one path for the current to take. These circuits are called **series circuits**. The charge must pass through all of the components in turn to return to the battery. This means that there is the same current in all of the components including the battery.

Parallel circuits contain branches where the current can divide and take different paths back to the battery.

Practical activity   Constructing circuits

Construct five circuits that match these circuit diagrams **a** to **e** below.

- Sketch your own diagrams showing what the circuits look like.
- Draw the circuit diagrams alongside your sketches.
- Which of the circuits are parallel circuits?

a
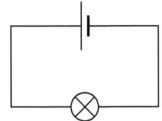

b
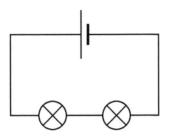

c
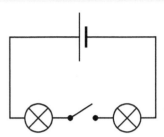

d
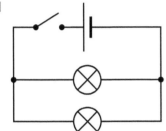

e
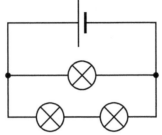

Key terms

- **circuit diagram**
- **cell**
- **series circuit**
- **parallel circuit**

Summary questions

1 Draw the correct circuit diagrams for these two circuits:

a
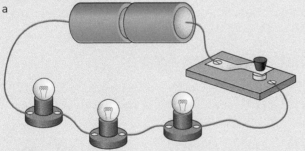

b
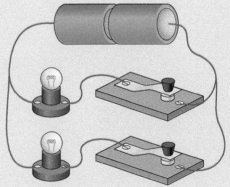

2 A student has drawn a circuit diagram poorly. Describe and correct the mistakes the student has made.

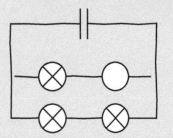

ammeter	—(A)—
voltmeter	—(V)—

Symbols for an ammeter and a voltmeter

Current

The current in a circuit is measured in amperes (A). The size of the current is measured with a device called an **ammeter**. Some ammeters have digital displays. They display a number to represent the size of the current. Some ammeters have analogue displays; a pointer moves and points to a number on a scale to indicate the size of the current.

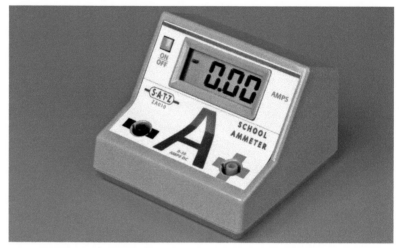

An ammeter is used to measure the size of a current

To measure the current, the ammeter has to be connected so that there is a current in it. This means the ammeter is placed in series with components.

The current is a measure of the amount of charge passing a point in the circuit each second. Large currents transfer more charge per second than small currents.

Potential difference (p.d.)

The potential difference across a device is measured in volts (V). The size of the p.d. is measured with a **voltmeter**. Voltmeters may have digital or analogue displays.

The voltmeter must be connected in parallel with the component across which the p.d. is being measured.

A voltmeter is used to measure the size of a potential difference

The potential difference across a component is a measure of how much energy is being delivered to that component by every unit of charge that passes through. The p.d. across a cell is a measurement of how much energy is being provided to each unit of charge passing through it.

Practical activity   Measuring current and potential difference in a circuit

Construct a simple circuit with a single lamp and battery. Use an ammeter to measure the current in the lamp and a voltmeter to measure the potential difference across the lamp.

Number of lamps	Current / A	p.d. / V
1		
2		
3		

Add a second lamp in the circuit directly after the first and retake the measurements.

Add a third lamp and retake the measurements.

- What happens to the current in the circuit when more lamps are added?
- What happens to the potential difference across each lamp?

 Do not use large currents.

Resistance

As more lamps are added to a simple circuit the current decreases. It becomes more difficult for the current to travel because the lamps have a **resistance** to the current.

All components resist the current to some extent, even the metal wires. Lamps have a much higher resistance than the wires and the current causes the filament inside the lamp to heat up. It is this heating effect that causes the lamp to glow.

Thick metal wires have a low electrical resistance. The wires transfer electrical energy efficiently. Cables used on pylons and underground are very thick because they need to transfer large amounts of electrical energy efficiently.

Key terms

- **ammeter**
- **voltmeter**
- **resistance**

Practical activity   Resistance and current

'The larger the resistance of a wire, the smaller the current in the wire will be for a fixed p.d.'

Design and carry out an investigation allowing you to produce a graph comparing the length of a wire and the current in the wire.

You will need to keep the potential difference across the wire the same throughout the experiment.

 Do not use large currents. Do not start practical work before your teacher has checked your plan.

Summary questions

1 A student measures the current in an increasing set of lamps. Plot a graph of this information and describe the pattern.

Number of lamps	1	2	3	4	5
Current /A	1	0.45	0.26	0.2	0.15

2 Copy this circuit and add a voltmeter and ammeter so that the current in the cell and the potential difference across the cell can be measured.

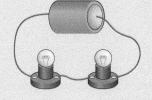

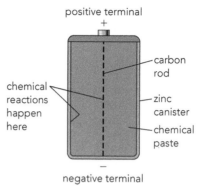

A simple cell

Cells produce a potential difference which causes a current in a complete circuit. A chemical reaction inside the cell causes the potential difference between the terminals. This p.d. drives the current. The current then provides energy to the components in the circuit as it travels.

chemical energy in cell → electrical energy

Practical activity   Lamp light

Build a simple circuit containing one cell and a lamp.

Add a second cell and note the change in brightness of the lamp.

Add a third cell.

- What does the brightness of the lamp show about the current in the circuit? Use an ammeter to check your answer.

- What would happen to the brightness of the lamp if the third cell has been added the wrong way around in the circuit?

Batteries and cells are available in a very wide range of shapes and sizes

Cells and batteries

Batteries are composed of a series of electrical cells connected together. Each cell produces a small potential difference. When cells are placed in series, the potential differences combine together and will produce a larger current in the circuit.

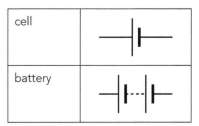

cell	
battery	

A cell and battery

Practical activity	Producing potential differences with fruit

Fruits contain acid and this acid can react with combinations of metals to produce a potential difference. Two different metal electrodes need to be placed into the fruit for a p.d. to be produced. You have been provided with some samples of fruit and a range of metal electrodes to place in the fruit. Design an investigation to test which combination of fruit and metals produce the greatest p.d.

Design a table to record your results clearly.

- You should plan to investigate the two important factors (choice of fruit and choice of metals) separately.

- Use the conclusions found from investigating both factors to decide on which overall combination of fruit and metals will produce the largest p.d.

 **Clear up any spilled fruit juice immediately.
Do not eat the fruit.
Do not start any practical work until your teacher has checked your plan.**

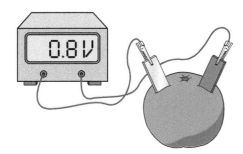

Expert tips

Try to remember the difference between a battery and a cell. A battery is a collection of cells working together.

 IGCSE Link...
You will study the concept of potential difference in detail during the IGCSE course.

Key terms

- **battery**

Summary questions

1 A lead-acid cell used in a car battery produces a p.d. of 2.1 V. The starter motor requires a p.d. of at least 12 V. How many cells need to be connected in series to provide this p.d?

2 A cell can be made by dipping two pieces of different metals into a beaker containing acid. Design a plan to investigate whether the potential difference produced depends on the surface area of the metal dipped in the acid.

Modelling current and potential difference

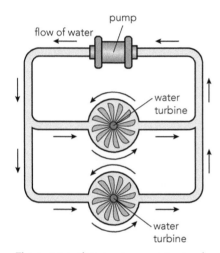

flow of water
pump
water turbine
water turbine

The pump pushes a water current around the circuit in a similar way to a cell pushing an electrical current. This shows a model of a parallel circuit

Using a model

It can be difficult to describe and understand the processes happening in a circuit. One way to help is to think about a model which is similar. One example is the water model:

- The movement of water represents the current in a circuit.
- A pump is used to represent the cells or battery – the pump pushes the water molecules just as the cell causes a push on the electrons.

In the diagram, a water pump pushes water around a circuit of pipes. The water current divides at the first junction and travels along two separate branches. The water in each branch then turns a turbine; transferring some energy. At the second junction the two currents merge and the water returns to the pump ready to travel around the pipes again.

Series circuits

In a water pipe system with only one route for the water to take, the same volume of water must pass each point in the pipe each second. This means that the current is the same throughout the pipe circuit.

This is similar to a series circuit where there is also only one path for the current to take. The same number of electrons (and amount of charge) passes through each component per second. This means the same sized current is in each component.

The water in the pipes is provided with a certain amount of energy by the pump when it passes through. The water then delivers some of this energy to each turbine as it passes through before returning to the pump.

Similarly, the electrons in a circuit are provided with energy by the cell and then deliver that energy as they travel through the circuit. The electrons can only deliver the same amount of energy they are provided with. This means that the total potential difference across the components in a series circuit is the same as the potential difference across the power supply.

Parallel circuits

In a water pipe circuit with a branch, the water would divide at the junction in the pipes and merge back together when the pipes re-join. The water going along each path would lose the same amount of energy before re-joining.

In a parallel circuit there are also junctions and the current can divide and take different routes. The current simply divides and later re-joins at the end of the branch. The total current leaving the cell is always the same as the total current returning to the cell.

	Series circuits	Parallel circuits
Paths	Only one path around the circuit.	More than one path around the circuit.
Current	Current is the same size in all parts of a series circuit including the cell or power supply.	The current can be different in each branch of the circuit. The current into any branch is the same as the current out of any branch.
Potential difference	The potential difference across each component can be different. The total potential difference across the components is the same as the potential difference across the power supply or cell.	The potential difference across each branch of the circuit is the same.

Circuit rules

Practical activity   Measuring the current and potential difference in circuits

Construct the circuits shown in each of the diagrams.

Measure the current and potential differences indicated.

As you have only been provided with one ammeter and one voltmeter, you will have to move the meters to different positions in the circuits to take all of the required measurements.

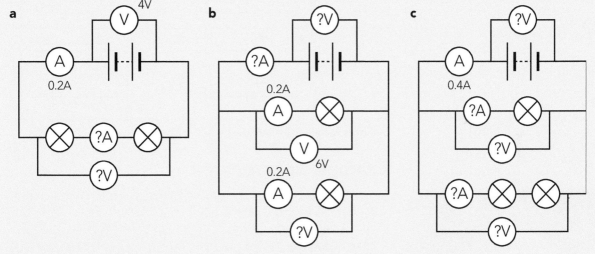

- Do your results match the rules about series and parallel circuits shown in the table?
- Explain why differences could arise.

Summary questions

1 In a water pipe model for a circuit:
   a) What represents the electrons?
   b) What represents the cell?
   c) What represents the components?

2 The water pipe model cannot fully describe an electrical circuit. Compare what happens when there is a break in a circuit and a break in a water pipe.

There is a much wider range of electrical components than just lamps and switches. Each component transforms electrical energy in a useful way and combinations of components allow circuits to provide a wide range of functions.

Practical activity   Exploring components

You have been provided with a range of circuits containing some of the components described below.

Test the circuits by operating the switches and observe the behaviour of the components.

- Describe what you see and what you think the circuits could be used for.

buzzer		LED	
electric bell		resistor	
motor	M	heater	
diode		fuse	

Symbols for electrical components

Example components

- Buzzers: buzzers produce simple sounds when there is a current in them. A buzzer is fairly small and is often used to alert a user that a device is on or that a button has been pressed.
- Electric bells: bells can produce louder noises than buzzers and so they can be used in alarms.
- Motors: motors transform electrical energy into kinetic energy allowing objects to move. For example, remote controlled cars have motors that operate their wheels and mobile phones have motors to allow them to vibrate when a message is received.
- Diode: diodes are devices that allow an electric current in one direction only. Some electrical components can be damaged if the current travels in the wrong direction. So diodes are used to protect these other components. The triangle on the diode symbol points in the direction of the allowed current.

A fire alarm bell needs to be very loud

- Light Emitting Diode: light emitting diodes produce light when there is a current in them. A LED will only allow current in one direction; in the same way as a normal diode. LEDs are much more efficient than lamps at transforming energy. Many modern torches use groups of LEDs as light sources.

- Resistors: a resistor can be used to control the size of the current in a circuit. There are many different types of resistors which can be used in circuits. Some resistors allow fixed currents in a branch of the circuit. Other resistors can be adjusted to control the current. A variable resistor can be used to control the current and so the brightness of a bulb.

- Heater: a heater is a simple device that transforms electrical energy into thermal energy. The power of a heater depends on the size of the current in it and the p.d. across it.

- Fuse: fuses are safety devices that prevent damage to other components in a circuit. Large electrical current can cause wires or components to overheat. This overheating could cause a fire. Fuses are designed to melt and create a gap in the circuit if the current becomes too large. The gap cuts off the current and the circuit stops operating before further damage is caused. Most mains operated devices have fuses.

> **IGCSE Link...**
> In the IGCSE Physics course you will build more advanced circuits and learn the symbols for some additional components.

Practical activity Design circuits

- Design and construct circuits which match these descriptions:

A circuit with a lamp and a buzzer that can be switched on independently.

A circuit that controls a motor that also has a buzzer that sounds whenever the motor is running.

A circuit with a heater that has a warning LED to indicate the heater is working.

A circuit with a lamp that will only operate when the battery is connected one way around.

Summary questions

1 Select an appropriate electrical component for each of these tasks:
   a) to heat a metal block   b) to make an efficient torch   c) for use in a door bell.

2 Describe what will happen in this circuit when each of the individual switches is closed.

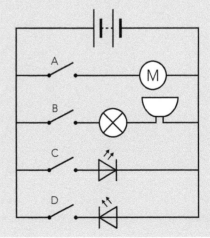

Sensing and responding to the environment

After this topic you should be able to:

- describe a range of sensors which can be used to detect environmental conditions
- suggest some uses for these sensors.

Circuits can sense the environment and respond to it. A **sensor** is a device that responds to changes in the environment and alters the behaviour of a circuit in response to the change.

Simple sensors

There are a wide range of simple sensors that can detect environmental conditions such as room temperature. Examples include:

- Thermistors: a thermistor is a device which can be used to measure the temperature.
- Light dependent resistor (LDR): a LDR can be used to detect light levels. Some LDRs detect visible light while others can detect infra-red light.
- Microphone: microphones detect sounds and produce electrical signals.
- A reed switch: a reed switch is a switch that can be operated magnetically. A small metal connection can be closed to turn on a circuit when a magnet is placed nearby.
- Pressure sensors: pressure sensors can detect forces acting on surfaces.
- Humidity sensors: humidity sensors can detect moisture levels.
- pH sensors: pH sensors can detect the pH of solutions by measuring the conductivity of the solution.

Digital sensors

Digital sensors collect information and convert it into digital form. Digital information is a series of low level (zero) and high level (one) signals which represent numbers. This digital information can be easily transmitted over long distances and processed by computer systems.

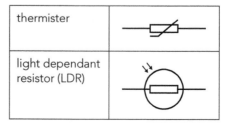

This digital thermometer converts the temperature measurement into digital form for processing. A microchip analyses the result and displays a message on the screen.

thermister	
light dependant resistor (LDR)	

Some sensor circuit symbols

Automatic circuits

Many circuits operate automatically, without the need for human input. An input sensor detects a change in the environment and produces a change in the circuit the sensor is part of. The change in the circuit is processed and this causes an output device to be switched on or off.

 IGCSE Link...
You will find out much more about the processing circuits that connect input sensors to output devices in the IGCSE Physics course.

Input		**Processing**		**Output**
• A sensor detects the environment		• The circuit uses the input to control an output device.		• The circuit activates a device in response to the input

- For example, street lamps may automatically activate when darkness falls:

Input		**Processing**		**Output**
• A light dependant resistor is used to detect it is dark.		• A circuit detects the input and is used to turn on a linked mains power circuit.		• A street lamp is switched on.

A modern smartphone has many input sensors and a range of outputs

Key terms

- **sensor**
- **digital sensor**

Summary questions

1. Which input sensors and output devices would you use in these situations? Draw a simple flowchart which describes the operation of a circuit that would:
   a) warm up a room when the room becomes cold
   b) turn on a fan when the air becomes too damp
   c) activate a sprinkler system when a fire is detected
   d) activate an alarm when a box is opened
   e) trigger a silent alarm system when a library becomes too noisy.

2. What are the inputs of a smartphone? What kinds of outputs can it produce?

The steady current provided by a battery is called a **direct current**. Over time, the chemical reaction in the battery slows down. Once the chemical reaction inside the battery has finished, there in no longer a p.d. and no current can be provided. The size of the current provided by a battery is also limited by the chemical reaction; the longer a battery is used, the less current it can provide.

Generators

To produce larger currents and potential differences, a generator is needed. Generators produce a potential difference from the interaction between a rotating coil of wire and a magnetic field. This process is called electromagnetic induction. The potential difference produced by generators is not a constant value. This causes the current to vary in size and direction. This type of current is called an **alternating current**.

Practical activity   Demonstrating alternating and direct currents

Your teacher will use a cathode ray oscilloscope to show you the different wave forms produced by a battery and an a.c. generator.

The p.d. produced by a battery is at a constant level. This would provide a constant current in one direction.

The p.d. produced by a generator changes in a wave pattern. This would produce a current that varies in size and changes direction rapidly.

The faster the generator spins, the higher the frequency of the wave.

The faster the generator spins, the greater the amplitude.

 Mains electricity should not be used in this demonstration.

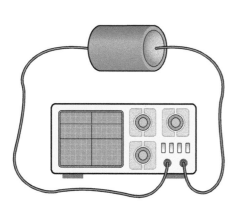

Oscilloscope trace for a direct current

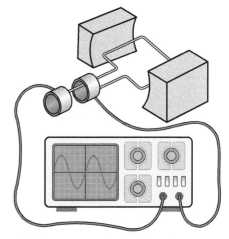

Oscilloscope trace for an alternating current

Generators can be rotated at different frequencies. The frequency of the alternating current is the same as the frequency of the generators' rotation.

Mains electricity

Mains electricity uses an alternating current. The potential difference of the mains supply and the frequency vary from country to country as shown in the table.

Example region or country	Mains voltage and frequency
Australia	230 V, 50 Hz
Saudi Arabia	127 V or 220 V, 60 Hz
North America	120 V, 60 Hz
United Arab Emirates	220 V, 50 Hz
Most of Europe	230 V, 50 Hz

The mains voltages in various countries

Dangers of mains electricity

Mains electricity can be very dangerous when used carelessly.

- The p.d. of mains electricity is high enough to force a current in a human's body. This current can cause burns or even death.
- Mains electricity can produce large currents. These large currents heat wires and can cause fires.

These electrical transformers have overloaded and a fire is starting. Suitable fuses would prevent this accident.

 IGCSE Link...
You will find out the details of electromagnetic induction in the IGCSE physics course.

Key terms

- **direct current**
- **alternating current**

Summary questions

1 Describe the differences between an alternating current and a direct current.

2 In which of these countries are the electrical generators rotating quicker? Saudi Arabia or the United Arab Emirates? Explain your answer.

3 Why is mains electricity more dangerous than electricity produced by a battery pack?

1 Copy this table adding the missing circuit symbols and component names.

Component	Symbol	Component	Symbol
Lamp			⊣⊢⊢
Ammeter		Switch	
	—(V)—		—(M)—

[6]

2 Complete the sentences by selecting the correct word:

A material that allows charges to pass through easily is a conductor/insulator. Materials that prevent charges passing through are are conductors/insulators.

Which of these materials conduct electricity?

a Copper
b Iron
c Glass
d Graphite
e Gold, wood, plastic
f Wood
g Plastic
h Sodium [4]

3 Complete this table showing how current and potential difference are measured:

Quantity	Current	Potential difference
Unit		
Unit symbol		
Meter used		
Meter placed in series or parallel?		

[4]

4 What are the readings shown on these analogue ammeters and voltmeters? [4]

a

b
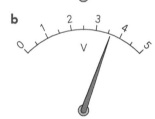

c

d

5 The current and potential differences were measured at various points for the following circuits. What are the missing values (**X**, **Y**, **Z**) for the ammeters and voltmeters? [3]

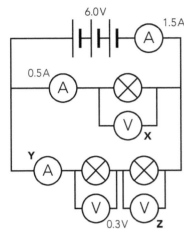

6 A child slides along a long dry slide at a fairground. The child gains electrons from the surface of the slide as she travels.

  a Why type of charge does the child gain? [1]

  b What type of charge does the surface of the slide gain? [1]

  c The child touches a metal surface and feels a small electric shock. Describe what happens to the electrons during this shock. [1]

7 Two students measured the effect that the length of a wire had on the size of a current in it. They placed the wire into a series circuit along with a 3.0 V battery and measured the current that in different lengths of the wire. Their results are shown in the graph.

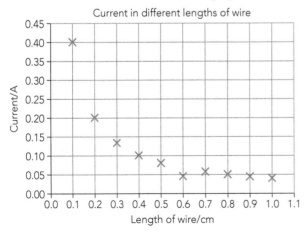

a Which result is anomalous? [1]

b Describe the relationship between the length of the wire and the current in it. [2]

c In what way would the graph be different if a battery with a larger p.d. had been used? [3]

8 A student was asked to investigate the current in two different components (**X** and **Y**) hidden in a box when a different potential difference was applied across them. The results are shown in the table below.

  a Which of the results are anomalous? [1]

  b Plot a graph showing the p.d. across the current in the two components **X** and **Y**. [4]

  c Use the graph to predict the current in the two components when there is a p.d. of 5.0 V across each of them. [2]

Potential difference / V	Current in component X / A	Current in component Y / A
1.0	0.20	0.50
1.5	0.40	0.95
2.0	0.60	1.40
2.5	0.90	1.70
3.0	1.00	2.10
3.5	1.20	2.35
4.0	1.40	2.50

Science *in context!*

From the edge of space to the deepest ocean

The conditions on the surface of the Earth can vary a great deal; from scorching deserts to icy wastes. Humans have managed to adapt to most of these conditions. However, our bodies can't survive in some of the extreme environments that we are beginning to explore. Scientists and engineers are designing the technology that will allow us to visit places that are totally inhospitable to humans.

A great fall

Skydiving from the edge of space

In October 2012, Felix Baumgartner jumped out of a capsule from a height of over 39 km and fell to Earth. The fall lasted over eight minutes before he reached the ground. For part of the journey he travelled faster than the speed of sound.

The technical challenges of performing this skydive were huge:

- At this height the atmosphere is very thin. So, designing a helium balloon and capsule large enough to lift a man but with a low enough density to reach this height was difficult.

- The air temperature 39 km from the surface of the Earth is −25 °C and during the ascent it dropped to under −60 °C. Felix's suit had to be designed to insulate him from these extremely cold temperatures.

- Air pressure is very low at these extreme heights. It is less than 1/100th of the air pressure acting on the surface of the Earth. Felix needed a pressurised suit to keep his blood within his body. If there has been a small tear in the suit, the air would have been forced out and he would have suffocated.

Plunging to the depths of the oceans

The DCV1 Deepsea challenger submarine

In March 2012, the film director James Cameron journeyed into the deepest part of the ocean, the Mariana Trench. Inside a small submarine James travelled 11 km beneath the surface of the ocean. He collected samples and photographed the strange creatures that live at this great depth. The conditions at this depth make it extremely difficult to visit. This was only the second time anybody had been this deep underwater.

- The submarine was designed to resist the massive pressures produced by the weight of 11 km of water pressing down on it. This pressure was over one thousand times the pressure at the surface of the ocean. A tiny design flaw in the submarine would lead to it being crushed.

- Thick steel walls and foam insulation meant that James was squashed into a space only 100 cm wide for many hours.
- The submarine had tiny windows many centimetres thick so that the huge compressive forces experienced would not break them. Many specially designed external cameras and lights were used to allow images to be captured.
- To sink into the water, the density of the submarine had to be greater than the density of the surrounding water. To return to the surface, a set of masses was released to decrease the submarine's density and make it buoyant.
- Although DCV1 is quite small it has a mass of over 11 000 kg due to the dense steel used to build it. To lift the DCV1 out of the water the control ship uses a powerful hydraulic crane. Once lifted out of the water the weight of the submarine causes the control ship to tilt and so the submarine has to be brought close to the ship quickly to reduce this tilting effect.

In this chapter you will explore how forces can be used to move objects and make them rotate. You will also find out about the density of fluids, such as water and air, and what causes pressure within them.

Key points

- Simple levers can be used to increase the size of forces used to move objects.
- Pulley systems can be used to reduce the size of forces needed to lift objects.
- The work done (energy transferred) by a force can be found using:

work done = force × distance moved in the direction of the force

- The density of a material is the mass per unit volume and can be found using:

density = mass / volume

- The pressure caused by a force on a surface is given by:

pressure = force / area

- Pressure in liquids increases with depth.
- Liquids are nearly incompressible and this allows them to be used in hydraulic machines as force multipliers.
- A force acting a distance from a pivot causes a turning effect called a moment.
- The size of a moment is given by:

moment = force × perpendicular distance from the pivot

Forces can cause objects to change shape or accelerate; but forces also transfer and transform energy when the force acts.

Forces in action: doing work

When a force causes an object to move there is a transfer of energy. For example, when a wooden block is pushed across a desk the frictional forces cause the desk and block to heat up. The transfer of energy caused by a force is called '**work**'.

The amount of work done by a force depends on the size of the force and the distance the force acts over (how far the object moves).

Work done (J)
= force (N) × distance moved in the direction of the force (m)

A large force acting over a large distance will perform a large amount of work.

Lifting with pulley systems

A pulley is one of the simplest machines available. A rope placed through a pulley allows us to change the direction of a force. Pulling downwards on a pulley system allows an object to be pulled upwards. If a single pulley is used, the upwards force is the same size as the downward force.

Combinations of pulleys

Two pulleys can be arranged to work together. This combination reduces the force needed to lift an object. The force needed to lift the object will be half the weight of the object but the rope will need to be pulled twice as far.

Pulley combinations are used on boats to generate large forces

IGCSE Link...
You will find out much more about 'work done' and energy transfer during the IGCSE Physics course.

Practical activity   Lifting loads with pulleys

Use a set of pulleys to investigate the force required to lift an object.

Use a newtonmeter to measure the weight of a load (a simple metal block).

Use a single pulley to lift the load.

- Record the force required, the distance the rope moved and the distance the mass moved.

Use a two pulley system to lift the mass.

- Record the force required, the distance the rope moved and the distance the load moved.

If you have time repeat the experiment with a three pulley system.

- Calculate the work done on the load using its weight and the height it was lifted through for each of the experiments.
- Calculate the work done pulling the rope using the force acting on the rope and the distance the rope moved.
- Compare these two values using the principle of conservation of energy.

	Work done on the load (= weight × change in height)	**Work done on rope** (= force × distance rope moves)
One pulley		
Two pulley		
Three pulleys		

 Do not lift large loads.

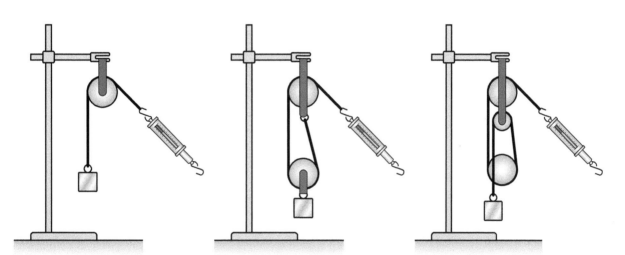

Three pulley systems. Using more pulleys reduces the forces needed to lift objects

Additional pulleys can be added to the system. This will reduce the force needed to lift objects but the rope will need to be pulled through a greater distance and so the same amount of work is done.

Key terms

- **work**

Summary questions

1 Explain why pulley systems are used to lift objects during the construction of buildings.

2 Which of these transfers the most energy? Show your working out.
   a) A car being pushed through 100 m using a force of 700 N.
   b) A motorcycle pulled through a distance of 400 m using a force of 150 N.

Levers, forces and work

Using levers to help

A lever is a simple machine that can be used to increase the size of forces. Levers are simple, stiff bars that can be rotated about a point called the **pivot** or fulcrum. A force applied a long distance from the pivot will produce a much larger force than if applied nearer to the pivot.

Changing force with levers

Force multipliers

A lever that is designed to produce increased forces is called a **force multiplier**. Most levers are force multipliers.

A crowbar is a solid metal bar that can be used to produce very large forces. The bar rotates about a pivot. A small force is applied on the long part of the crowbar, a large distance from the pivot. This small force produces a large force on the other end of the bar, which is very close to the pivot point.

A wheelbarrow pivots around its front wheel. The handles of the wheelbarrow are further away from the pivot than the load. This means that a smaller force is needed to lift the objects in the wheelbarrow.

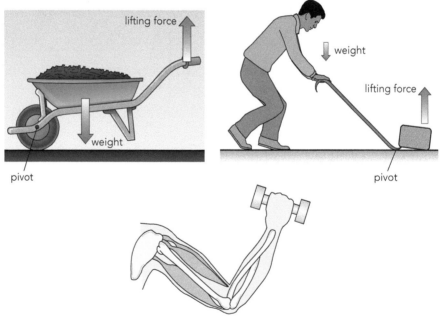

Levers can be force multipliers or distance multipliers

Distance multipliers

Some levers produce decreased forces but increase the distance an object moves. The muscles in an arm contract only a few centimetres but this makes the end of the arm (hand) move through a large distance.

Doing work with levers

Levers and the law of conservation of energy

The law of conservation of energy shows that the amount of work done (energy transferred) by a lever cannot be greater than the work done by the pushing force. This means that to produce a large force from a small force, the small force has to move through a greater distance. For example, when a crowbar is used the handle is moved through a large distance, while the claw end only moves through a small distance.

Practical activity   Testing some levers

You have been given a range of levers. Test the levers to see how they operate. You should identify the pivot on each lever.

Use a claw hammer to remove a nail from a block of wood.

Use a crowbar to separate two blocks that have been nailed together.

Use a spanner to tighten and loosen a bolt.

Use some scissors to cut through paper.

 Do not use large forces in the experiment.

Key terms

- **pivot**
- **force multiplier**

Summary questions

1 This diagram shows a screwdriver being used to lift off the lid from a tin of paint.
Copy the diagram and add labels to show where the forces acting on the screwdriver and lid would act. Label the pivot.

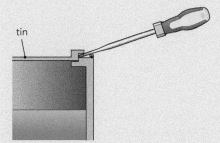

tin

2 What effect does increasing the length of a lever have on the size of the force the lever can produce?

3 Why is it easier to cut card with large scissors than with small scissors?

2.3 Moments

Learning outcomes

After this topic you should be able to:

- describe the turning effect caused by a force
- calculate the moment due to a force.

Calculating moments

The turning effect of a force is called a **moment**. The size of a moment depends on the size of the force and the perpendicular distance that the force is from the pivot:

moment (Nm) = force (N) × perpendicular distance from pivot (m)

The unit of a moment is the newton metre (Nm).

Example

A force of 20 N is applied 30 cm from the pivot of a crowbar. What size moment will this produce?

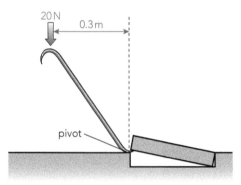

moment = force × perpendicular distance from pivot
= 20 N × 0.3 m
= 6.0 Nm

Expert tips

It is best to convert any lengths from centimetres into metres before calculating moments as shown here when the distance of 30 cm has been converted to 0.3 m.

Describing the directions

Because moments cause rotation, the direction they act needs to be stated. It is described using the directions a clock's hands move. Clockwise moments will cause an object to rotate clockwise around a pivot. Anticlockwise moments will cause the object to rotate anticlockwise.

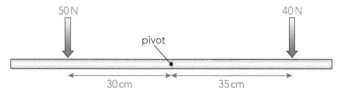

Moments acting in different directions

In the diagram above, the 40 N force is trying to make the bar move clockwise while the 50 N force is trying to make the bar move anticlockwise. The bar will begin to rotate in the direction of greater overall moment.

Adding moments

Moments can be added in a similar way to forces to find the resultant moment acting on an object. The moments are calculated individually and then added together.

Example

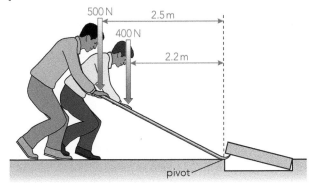

Two workers are using a large lever to lift up a concrete slab. Worker A applies a force of 400 N at a distance 2.2 m from the pivot. Worker B applies force of 500 N at a distance 2.5 m from the pivot. What is the total turning effect produced?

Total moment = moment provided by worker A + moment provided by worker B

Total moment = (400 × 2.2) + (500 × 2.5)

Total moment = 880 + 1250

Total moment = 2130 Nm

Moments acting against each other can be subtracted from each other in a similar way.

Key terms

- **moment**

Summary questions

1 Calculate the clockwise and anticlockwise moments caused by the forces acting on the metal rod in the diagram above.

2 A screwdriver is used to help remove the lid off of a tin of paint. A force of 10 N is applied to the end of the screwdriver at a distance of 15 cm from the pivot point.
   a) What is the moment of the force?
   b) Is the screwdriver acting as a force multiplier or a distance multiplier?

3 Calculate the missing values from this table:

Force /N	Distance from pivot /m	Moment /Nm
20	0.3	
3	1.2	
400		5000

Equilibrium

In many situations the moments on both sides of a pivot are equal to each other. When this happens the turning effects cancel each other out.

An object is said to be in **equilibrium** when there is no overall moment and no resultant force acting on it. An object in equilibrium is not accelerating or changing its rotational speed.

The principle of moments

The principle of moments states that:

For an object in equilibrium, the moments acting clockwise are equal to the moments acting anticlockwise.

Practical activity   Verifying the principle of moments

Use a stiff ruler, a triangular pivot and some small masses to verify the principle of moments. Remember 100 g has a weight of 1.0 N.

- Balance the ruler on the triangular pivot. If the ruler is uniform (the same all the way along) then the ruler should balance at the central point.
- Place a 25 g mass 20 cm from the pivot on the left side of the ruler.
- Place a 50 g mass on the right hand side of the ruler and adjust the position of the mass until the ruler balances. Record this position.
- Repeat the process using different masses at different distances. Copy and complete the table below.

Anticlockwise (left-hand side)			Clockwise (right-hand side)		
Force / N	Distance / m	Moment / Nm	Force / N	Distance / m	Moment / Nm
0.25 N	0.20	0.05			

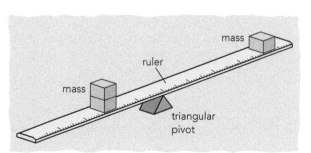

Solving moment problems

The principle of moments can be used to analyse situations where objects are in equilibrium.

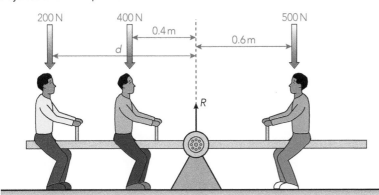

In this diagram the seesaw is in equilibrium. This means that:

- there is no overall resultant force.
- the clockwise moments = the anticlockwise moments.

These two facts can be used to find the values of the missing force and missing distance.

As there is no resultant force, the upwards force R must be equal to the total of the downward forces:

$R = 400\,N + 200\,N + 500\,N = 1100\,N$

The moments are also balanced:

Clockwise moments = anticlockwise moments
$$(0.6 \times 500) = (0.4 \times 400) + (d \times 200)$$
$$300 = 160 + 200d$$
$$200d = 300 - 160$$
$$d = 140 / 200$$
$$d = 0.7\,m$$

Expert tips

During moment calculations it can help if you put each moment inside a set of brackets as shown in the example.

IGCSE Link...

You will solve more problems using moments during the IGSE Physics course.

Key terms

- **equilibrium**

Summary questions

1 What are the two conditions required for an object to be in equilibrium?

2 Three people of equal weight sit on a seesaw of total length 4.0 m. Sketch a diagram showing the possible positions where the three people could sit so that the seesaw would be balanced.

3 A crane is in equilibrium when it holds a load of 50 kN on its cable 15 metres from the pivot. The crane uses a large block of concrete weighing 90 kN to balance the load. How far should the concrete block be placed from the pivot?

When a force is applied to an object, the force acts over an area. For example when you push a trolley, the force you use acts over the area of contact between your hands and the trolley.

Defining pressure

Pressure is defined as the force acting per unit area. This can be shown in the equation below:

$$\text{pressure (N/m}^2\text{)} = \frac{\text{force (N)}}{\text{area (m}^2\text{)}}$$

The unit of pressure, newtons per square metre (N/m^2) is also called a pascal (Pa).

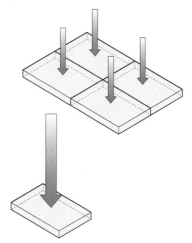

When a force acts over an area, a pressure is produced

Measuring the pressure on a surface

To find the pressure on a surface, the force acting on the surface and the area the force acts on need to be known.

Example

A box weighing 500 N is placed on a desk. The area of the bottom surface of the box is 0.25 m^2. What pressure does the box exert on the desk?

Pressure (N/m^2) = force (N) / area (m^2)
= 500 N / 0.25 m^2
= 2000 N/m^2

The effects of pressure on surfaces

Objects with small areas of contact can produce very high pressures that cut into surfaces:

- A nail has a very small area of contact with a piece of wood. As the nail is hit by a hammer the nail will cut into the wood easily because the force is concentrated onto this small area of contact.

- Knives have sharp blades with cutting edges that have very small surface areas. The small area means that only a small force needs to be used to push the knife into materials.

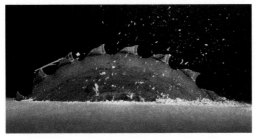

This circular saw blade cuts through wood easily

Some objects have large areas of contact and so do not damage surfaces:

- Large trucks have many wide wheels so that the area of contact on the road is large. This reduces the pressure on the road so that it does not become worn and damaged.
- A camel has pads on the bottom of its feet that spread out when it walks on sand. This increases the area that the camel's weight acts on and prevents the camel sinking into the sand.

The camel's weight is spread to reduce the pressure on the sand

Practical activity   Your pressure on the floor

You can find the pressure that you produce on the floor by measuring your weight and the area of contact between you and the floor.

- Measure your weight with bathroom scales.
- Stand on squared paper and draw around your shoes. Count the squares and estimate the area of contact between your shoes and the floor.
- How much pressure do you exert on the floor?
- What happens to the pressure if you stand on one leg?

Measure the area of contact between the feet of your stool or chair and the floor. Use this to calculate the pressure acting on the floor when you sit down. (HINT: Don't forget about the weight of the chair.)

Expert tips

Be careful with the units of pressure – it can be confusing to try to convert from N/cm^2 to N/m^2 and so it is better to just use N/m^2.

🔗 *IGCSE Link...*
You will tackle more problems involving pressure in the IGSE Physics course.

Key terms

- **pressure**

Summary questions

1 What is the pressure caused on the ground by an elephant of weight 20 000 N if each foot has an area of 0.2 m²?

2 A drawing pin is pushed into a wooden block using a force of 15 N. The sharp end of the drawing pin has an area of 0.5 mm² (0.000 000 5 m²).
   a) What pressure does the pin exert on the wooden block as it is pushed in?
   b) Why isn't the thumb of the person pushing the pin into the block damaged by the pressure acting on it?

Gas pressure

After this topic you should be able to:

- explain why gases cause pressure on a surface
- describe how the pressure in a gas can be changed by altering the temperature or volume
- explain results using scientific ideas and communicate these clearly to others.

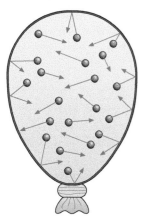

The pressure within the balloon is caused by the movement of the particles of air

The cause of gas pressure

In a gas, the particles are spaced far apart compared with particles in liquids and solids. The random and fast movement of the gas particles is the cause of pressure on the surfaces of containers holding the gas:

- The gas particles collide with surfaces and change direction.
- The change in direction is caused by forces acting between the gas particles and the surface.
- There are many collisions between the particles and the surface so there are many small forces acting over an area.
- Forces acting over an area cause a pressure.

Changing gas pressure

The temperature, volume and pressure of a gas are closely linked. For a fixed mass of gas:

- Reducing the volume of the gas increases the pressure of the gas. Reducing the volume will force the same number of particles into a smaller volume. More particles will be colliding with the container every second. This will increase the overall force on the container and therefore the pressure.
- Increasing the temperature of the gas increases the pressure. An increase in the temperature will make the particles of gas move more quickly. The gas particles will hit the container walls more often and with greater force. This increased frequency of harder collisions will increase the pressure of the gas.

Practical activity   Comparing a gas

Use a large plastic syringe to see how gases behave when they are compressed. The syringe has been sealed so that the gas cannot escape. You can squash the gas by pushing on the plunger and observe what happens to the volume.

You can also place different weights on the plunger to compress the gas.

Use the ideas you have seen to design an experiment to find out how the volume of a gas changes when it is compressed by an external force. The mass of the gas must remain constant.

- Explain the results of your experiment and share your conclusion with the rest of your class.

 Do not press the plunger so hard that you hurt yourself. Only use plastic syringes.

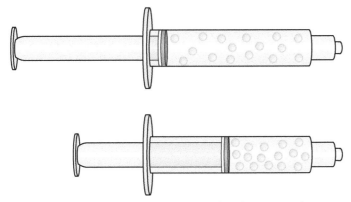

The particles in a gas are pushed closer together when the gas is under pressure

Pressure in the atmosphere

The Earth's atmosphere is made up of a mixture of gases that form a layer over 100 km high. On the Earth's surface the pressure caused by the atmosphere is approximately 100 000 N/m².

The higher up from sea level, the lower the pressure in the atmosphere becomes. This means that the density of the atmosphere decreases with height.

Pressurised cabins

Jet passenger planes can reach a height of 10 km above sea level. Atmospheric pressure is only 30 000 N/m² at this height. At this pressure it would be impossible to breathe in enough oxygen and so aeroplane cabins need to be pressured. The cabin pressure is kept higher than the external air pressure.

The cabin on a large airliner has to be pressurised to provide enough oxygen to the passengers

🔗 *IGCSE Link...*
You will find out a great deal more about the behaviour of gases in the IGSE Physics course including a simple way to measure the pressure of gases.

Summary questions

1 Copy and complete:

Compressing a fixed mass of gas into a smaller volume will _____ the pressure of the gas.

Decreasing the temperature of a fixed mass of gas while keeping the volume constant will _____ the pressure of the gas.

2 If a sealed can of compressed air is heated the can may explode. Explain why this might happen.

3 A vacuum pump can be used to remove all of the air from a metal can. Describe and explain what would happen to the can.

The cause of pressure in liquids

The pressure acting on an object in a liquid is caused by the weight of the liquid above it. The pressure increases with the depth because there is more weight from the liquid above acting downwards on the object.

For example, a submarine experiences pressure when it is underwater. This is explained below:

- The weight of the water above the submarine acts on the submarine.
- The weight acts over the surface of the submarine and so generates a pressure.
- As the submarine dives deeper there is more water above it and so the weight of water increases.
- There is a greater force acting on the same area of the submarine.
- The pressure acting on the submarine increases.

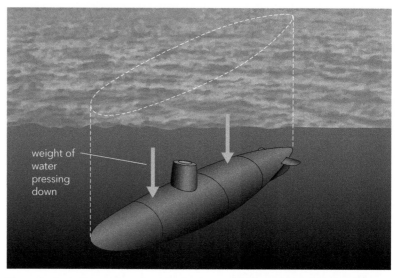

weight of water pressing down

The pressure acting on the submarine increases as it dives deeper into the ocean

Practical activity   Demonstrating increasing pressure in liquids

You have been given a container with small holes in the side.

Fill the container with water and observe what happens.

- Describe what happened to the water coming from the different holes.
- Compare how far the water travels with the depth of the liquid.
- Explain why the results show that the pressure in the water increases with depth.

 Carry out the experiment next to a sink and allow the water to spill into it.

Pressure beneath the ocean

Pressure changes with depth in the ocean. The pressure increases rapidly as the depth increases. Large pressures underwater make exploration difficult.

Deep sea diving

Diving to a depth of 10 m will double the pressure acting on a person to 200 000 N/m^2. This is twice the atmospheric pressure at sea level. This increased pressure will make it difficult for the diver to breathe as the pressure pushes against his or her chest. The pressure increases by an extra 100 000 N/m^2 with every additional 10 m of depth.

At depths of over 50 m nitrogen gas from the diver's air supply begins to dissolve in the diver's blood. This can cause the diver to fall unconscious. On returning to the surface the dissolved gases can cause the diver to suffer from 'the bends', which can be fatal. Divers have developed gas mixtures that replace the nitrogen in the air with helium, which is less harmful, but these can only work to depth of 500 m.

Submarines

Submarines can travel deeper into the oceans than divers. The rigid steel shell of a submarine allows the pressure inside the submarine to be much lower than the pressure outside. Because the atmosphere in a submarine is kept close to the surface pressure, the submarine can return to the surface quickly without causing harmful effects to the crew.

Passengers on this recreational submarine won't experience the change in pressure as they dive beneath the ocean

 IGCSE Link...
You will calculate the pressure of a liquid at different depths during the IGCSE Physics course.

Summary questions

1 Copy and complete:

The particles in a liquid are more _____ packed than the particles in a gas and so liquids are _____ dense.

As you dive deeper into a liquid the pressure _____. This is because the _____ of the water above you increases.

2 A gas-filled balloon is attached to a very heavy weight and dropped into an ocean. Describe, and explain, what would happen to the volume of the balloon and the pressure of the gas within it as the balloon sinks.

Hydraulic pistons transmit forces

It is very difficult to compress liquids inside a container because the particles are already closely packed together. If a force is applied to a liquid, a pressure is produced throughout the whole volume of the liquid.

Transmitting pressure

Hydraulic machines use the idea of transmitting pressure through liquids to transmit forces.

- A force is applied to one part of the hydraulic system.
- The force produces a pressure throughout the liquid.
- The pressure produces a force at the other end of the system.

If the areas that the force acts on at the different ends of the hydraulic system are different, then the forces will also be different sizes. This allows a hydraulic machine to act as a force multiplier or a distance multiplier.

Hydraulic machines operate using fluid filled pipes that can bend around corners. This means that the direction in which the force is acting can be changed easily.

Practical activity — Demonstrating the principle of hydraulic machines

Two syringes of different sizes have been connected by rubber tubing. The syringes contain water and this is difficult to compress.

Push on the smaller of the syringes and observe what happens to the larger syringe.

Push on the larger syringe and observe the smaller syringe.

- Design an experiment to show that applying a force on the smaller syringe will allow the larger syringe to produce a larger force.

 Do not push the syringes too hard; they might leak.

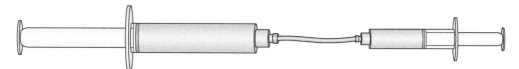

Gases entering a hydraulic system will cause the machine to stop functioning. The pressure in the liquid will fall as the gases become compressed and take up less volume. This means that the force cannot be transmitted effectively.

Conservation of energy

Although the forces produced by a hydraulic system can be larger than the input force, the work done (energy transferred) does not change. The smaller force will have to move through a larger distance. This means that hydraulic systems follow the law of conservation of energy in the same way that pulleys and levers do.

Hydraulic calculations

The pressure equation can be used to find the size of the forces involved in hydraulic machines.

Example

Two liquid syringes connected together

A syringe has an area of 5.0 cm² and a force of 25 N is applied to it. This force is transmitted through a liquid to another syringe with an area of 20 cm². Calculate the force produced by the second syringe.

1. Find the pressure acting on the first syringe:

   Pressure = force / area
   $$= 25\,N / 5.0\,cm^2$$
   $$= 5.0\,N / cm^2.$$

   This is the same pressure that will be acting on the second syringe. This pressure is used in the next part of the calculation:

2. Then find the force produced by the second syringe by rearranging the pressure equation:

   Force = pressure × area
   $$= 5.0\,N / cm^2 \times 20\,cm^2$$
   $$= 100\,N$$

 IGCSE Link...
The braking system of cars relies on hydraulics as you will find out in the IGCSE Physics course.

Key terms

• **hydraulic machines**

Summary questions

1 Why can't gases be used to transmit forces in the same way as liquids are used in hydraulic systems?

2 What advantages do hydraulic systems have when compared to simple levers or pulleys?

3 Two liquid filled syringes are connected together by sealed tubing. Syringe A has an area of 4.0 cm² and syringe B has an area of 1.5 cm². A force of 50 N is applied to syringe A.

   a) Calculate the pressure produced in the liquid.

   b) Calculate the force produced in syringe B.

After this topic you should be able to:

- define and calculate density
- measure the density of liquid samples
- explain how the density of gases can be found.

Defining density

The density of a material is defined by the equation:

$$\text{density (kg/m}^3\text{)} = \frac{\text{mass (kg)}}{\text{volume (m}^3\text{)}}$$

To measure density, both the mass and volume need to be measured. At Stage 7 you used this idea to measure the density of samples of solid materials. For example, you can calculate the density of iron, given a block of iron with a mass of 395 kg and a volume of 0.05 m².

Density of iron = 395 kg/0.05 m² = 7900 kg/m²

The density of liquids and gases can be measured in a similar way.

Measuring the density of liquids

A liquid will fill the bottom of a container and so a measuring cylinder or beaker can be used to find the volume. The mass can be measured using a balance.

Practical activity   Measuring the denisty of a liquid

Find the density of various liquids using the equipment provided. You will need to find the mass and the volume of samples of the liquids and then use the density equation. To produce an accurate measurement of density you will need to make sure that the mass and volume of the liquids are measured as accurately as possible.

When you measure the mass you will need to subtract the mass of any container that you use.

Use precise instruments to measure the mass and volume.

- What was the precision of the balance used in the experiment?
- How accurately did you measure the volume of the liquid?
- Why would using larger samples produce a more accurate measurement of density than using small samples?

 Clear up any spills of liquids immediately. Do not allow liquids to spill on any electronic balances.

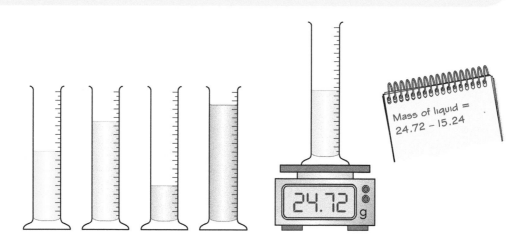

Mass of liquid = 24.72 – 15.24

24.72 g

Measuring the density of gases

To measure the density of a gas, the mass and volume are still used. The sample of gas can be pumped into a container of known volume. The mass of the gas and container is measured. The gas can then be removed by a vacuum pump and the mass of the empty container can be measured. The mass of the gas can then be calculated from the two measurements.

Example

A metal container of volume $1.5\,m^3$ has a mass of $20.0\,kg$ when full of methane gas. The container has a mass of $19.7\,kg$ when it is empty. What was the density of the gas when it was inside the container?

The mass of gas was $0.3\,kg$ ($20\,kg - 19.7\,kg$)

Density = mass / volume
$$= 0.3\,kg\ /\ 1.5\,m^3$$
$$= 0.2\ kg\ /\ m^3$$

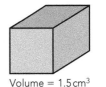

Volume = $1.5\,cm^3$
Empty mass = $19.7\,kg$
Full mass = $20\,kg$

Increasing density with depth

The density of a fluid is increased if there are forces that compress the fluid. The particles in the fluid are forced closer together and so there are more particles in the same volume.

In a liquid the particles are already closely packed and so very large compressive forces are needed to increase the density by a small amount.

In a gas there is much more space between the particles and so the particles can be forced closer together more easily. This means that it is easy to increase the density of a gas by squashing it.

Summary questions

1 Copy and complete this table to find the missing values for the liquids:

	Mass / g	Volume / cm³	Density / g/cm³
Mercury	67.5	5.0	
Seawater	140		1.03

2 A large empty balloon has a mass of $0.03\,kg$. When the balloon is filled with xenon gas, the mass of the balloon increases to $0.15\,kg$ and the volume is $2000\,cm^3$.

   a) What is the mass of the xenon gas in the balloon?

   b) Calculate the density of the xenon gas.

   The density of the air in the room is $0.0012\,g/cm^3$.

   c) Describe and explain what will happen to the balloon if it is released.

1 A spanner is used to untighten a nut as shown in the diagram.

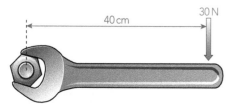

    40 cm   30 N

  a Calculate the turning effect of the spanner. *[2]*

  b Is a spanner a force multiplier or a distance multiplier? *[1]*

2 Three children are playing on a seesaw as shown in the table.

Child	Weight	Position compared to pivot
X	300 N	1.0 m to the right
Y	200 N	1.5 m to the right
Z	500 N	1.5 m to the left

  a Calculate the moments produced by the three children. *[3]*

  b Is the seesaw in equilibrium? *[1]*

3 A pupil weighing 400 N sits on a three-legged stool weighing 100 N. Each of the chair legs is in contact with the floor over a surface of 4 cm^2

  a What is the total weight of the pupil and stool? *[1]*

  b What is the total area on contact between the stool and the floor? *[1]*

  c What pressure does the chair exert on the floor when the pupil sits on it? *[1]*

4 A crane uses a system of four pulleys and a steel cable to lift a concrete block to the top of a tall building. The block weighs 5000 N and is lifted to a height of 60 m.

  a How much work needs to be done lifting the block? *[1]*

  b How much force needs to be applied to the cable to lift the block? *[1]*

  c How far does the cable need to be pulled in order to lift the block 60 m? *[1]*

5 Which of these factors will affect the pressure in a 1.0 kg sample of gas?

  A the temperature of the gas

  B the colour of the gas

  C the volume of the gas

  D the smell of the gas

  E the kinetic energy of the gas molecules. *[3]*

6 A hydraulic lifting platform is used to lift cars in a factory. The hydraulic system is made from two cylinders (X and Y) connected by a hydraulic pipe. Cylinder X has a cross sectional area of 0.2 cm^2. A force of 5000 N is applied on it.

  a What is the pressure generated in the hydraulic fluid by cylinder X? *[1]*

  Cylinder Y has a cross sectional area of 1.5 m^2.

  b What is the force produced by cylinder Y? *[1]*

7 Two students are investigating how the pressure of a liquid varies with depth using a tall measuring cylinder with a hole in the side 5 cm from the bottom. They fill the measuring cylinder with different depths of water and allow the water to come out of the hole. They measure the maximum distance the jet of water travels horizontally before hitting the desk.

Water depth / cm	5	10	15	20	25
Distance water travels / cm	4.8	8.6	12	14.3	15.2

  a Plot a graph to show the pattern in the data. *[4]*

  b Describe the pattern shown by the graph. *[1]*

  c Explain the way in which this evidence shows that the pressure of the water varies with the depth of the water. *[2]*

8 A team of research scientists is studying the freezing of water to help them understand why water in ponds freezes at the top before the water freezes at the bottom. They use a sample of water with a mass of 100.000 g and allow it to warm very slowly. The scientists

record the volume of the water very precisely as the temperature increases.

Temperature / °C	Volume / cm³	Density / g/cm³
0.0	100.035	
0.5	100.030	
1.0	100.025	
1.5	100.020	
2.0	100.015	
2.5	100.010	
3.0	100.005	

a Calculate the density of the sample of water at the different temperatures. [3]

b Is the water sample expanding or contracting as it warms up? [1]

c The density of ice at 0.0 °C is 0.9167 g/cm³. Why does the ice float on the water? [1]

9 A group of students is discussing the transfer of energy (work done) by simple machines. They make the following **incorrect** statements:

Student A: "A lever can produce a bigger force than you put in to it so it must make energy."

Student B: "So does a pulley system – you can lift heavy objects using forces smaller than the objects weight so it is doing more work than you put into it.".

a Which principle shows that these statements must be wrong? [1]

b State the formula for work done by a force. [1]

c Use the formula to explain the mistakes the students have made. [2]

Science *in context!*

Increasing demand for energy

The population of the Earth is growing. This increase in population means that there is a rapidly growing demand on the Earth's limited resources. As more countries become industrialised, the global energy demand increases because more people use electrical devices, drive cars, install air conditioning and use aeroplanes.

Over the past two centuries fossil fuels have been the main source of the energy we need. Billions of tonnes of coal, oil and natural gas are burned every year to produce our electricity. These non-renewable resources are becoming more expensive and more difficult to find and we cannot continue to rely on them.

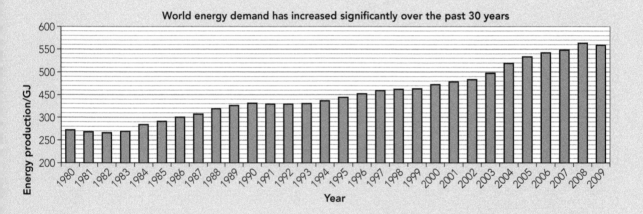

World energy demand has increased significantly over the past 30 years

The greenhouse effect

Carbon dioxide, methane and water vapour are greenhouse gases. Greenhouse gases prevent some of the thermal energy absorbed by the Earth from escaping back into space. Without this 'greenhouse effect' the Earth would not be warm enough for life to exist on it.

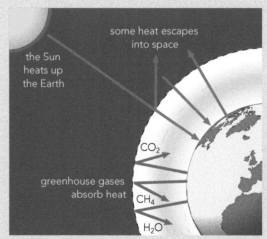

The greenhouse effect keeps the Earth warm by preventing some of the thermal energy from escaping into space

Global warming

When fossil fuels are burned in power stations or vehicles extra carbon dioxide is released into the atmosphere. The rise in the level of atmospheric carbon dioxide has been linked to increasing global temperatures and changes in weather patterns. The greenhouse effect has been increasing and additional thermal energy is being trapped by the gases. Global warming has become a major problem.

There are natural processes that can cause the Earth to warm up. However, most scientists are convinced that man-made global warming will be disastrous for humans if it is not reduced. Possible effects include:

- land becoming flooded as sea levels rise
- global food shortages as useful land is lost
- changes in weather patterns destroying crops or causing mass migration.

Changes in weather can lead to floods

Exploring new resources

A wide range of alternative sources of energy have been developed but none of these are as easy to use or as cheap as fossil fuels. Over the next century scientists and engineers will try to improve these alternative sources of energy and stop global warming before it causes too much damage.

In this chapter you will study how thermal energy is transferred and then look at a wide range of methods of producing electricity.

Key points

- Thermal energy can be transferred in four ways: conduction, convection, radiation and evaporation.
- Thermal conduction involves energy being passed from particle to particle.
- Thermal conduction is the only method of thermal energy transfer through solid materials.
- Convection involves the flow of particles due to the changes of density in a fluid (a liquid or a gas).
- Radiation involves the transfer of energy by infra-red radiation.
- Thermal energy transfer by radiation can happen through a vacuum.
- Fossil fuels are burnt to produce electricity in power stations.
- The energy from sunlight can be transformed into useful energy for heating or electrical energy.
- Wind turbines produce electricity from the kinetic energy of wind.
- Hydroelectric power stations produce electricity from the potential energy stored in water.
- Tidal power stations can produce electrical energy using the tidal flow of water.
- Wave driven generators can be used to produce electricity around coastlines.
- Nuclear power can produce large amounts of energy but there are significant risks that need to be taken into account.

Measuring temperature

The **temperature** of an object can be measured with a thermometer using a scale called the Celsius scale. This scale is based on the boiling point and freezing point of water.

-273 —— The lowest temperature possible: 'absolute zero'.

-89.2 —— The coldest temperature recorded at the surface of the Earth (Antarctica).

0 —— The freezing point of water.

37 —— The average temperature of a human body.

56.7 —— The highest temperature recorded at the surface of the Earth (Arizona).

100 —— The boiling point of water.

700–800 —— Bunsen burner flame.

Some important points on the Celsius (°C) scale

There are several different types of thermometer. The two most commonly used are described below:

- A 'liquid-in-glass' thermometer contains mercury or coloured alcohol. The liquid expands when the temperature increases. As the liquid expands up a thin tube inside the glass, the scale can be read to measure the temperature of the thermometer.

- A thermocouple is an electric junction. It produces a small potential difference that can be measured to give a temperature reading. Thermocouples are often used in electronic thermometers.

Glass thermometer

Thermocouples are often used in electronic thermometers

Temperature and energy

When an object is heated the temperature of that object increases. This is because the particles inside the object gain energy. Increasing the temperature of a larger object requires more energy than heating a smaller object made from the same material.

The **thermal energy** in an object depends on:

- the temperature of the object – the higher the temperature, the more thermal energy the object will contain
- the mass of the object – the larger the mass, the more thermal energy the object will contain
- the chemical composition and bonding in the object – some materials can store more energy than others.

Practical activity   Transferring energy to water

Heat three different volumes of water by providing them with the same amount of energy. You can provide this energy by using an electrical heating element or by using a Bunsen burner to heat the different volumes of water for the same amount of time. You may wish to perform a trial run of this experiment to establish how long to heat the water for.

 Do not let the water boil – select a suitably short heating time.

Make sure there is a measurably large change in temperature – select a suitably long heating time.

- Give a conclusion about the relationship between the rise in temperature and the mass of water heated.
- Compare your results with other students' results to establish the reliability (reproducibility) of the experiment.
- Was all of the energy transferred to the water? In what way was energy lost?

Volume of water / cm³	Mass of water / g	Start temperature / °C	End temperature / °C	Change in temperature / °C
50	50			
100	100			
150	150			

 Be careful with hot water and the flame from the Bunsen burner.

Key terms

- **temperature**
- **thermal energy**

Summary questions

1 A rock with a temperature of 300 °C is dropped into a large bath of water at room temperature. Explain why the temperature of the water only rises slightly.

2 Why does a large volume of water take longer to boil in a kettle than a small volume of water?

Learning outcomes

After this topic you should be able to:

- describe how thermal energy passes through solid materials by the process of conduction
- investigate the differences between a conductor and an insulator.

The process of conduction

Thermal energy can pass through solid materials by a process called **conduction**. The energy spreads slowly through materials.

- When one part of the material is heated the particles gain energy.
- The gain in energy causes the particles to vibrate about their positions more.
- The vibrating particles collide with nearby particles and pass some of the thermal energy on to them, causing these new particles to vibrate more vigorously.
- This process continues with the energy being slowly passed along until all parts of the object are heated.

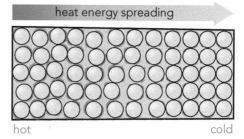

Thermal energy moves from the hot part of the material to the cold part

Metals are very good conductors of thermal energy. The energy can quickly be passed from one part of the metal to another by the free-moving electrons that are responsible for electrical currents.

Insulation

To prevent the flow of thermal energy by conduction, insulating materials can be used. Insulators are materials that are poor conductors of thermal energy.

Air is a very poor conductor and so materials that contain large amounts of trapped air, such as expanded polystyrene or wool, are effective insulating materials.

Expanded polystyrene can be used for cups containing hot liquid. It is such a poor conductor that the thermal energy cannot escape quickly. What could you do to keep the liquid in these cups hotter for longer?

Most cooking utensils are made from metal because metals are good thermal conductors with high melting points.

Place a metal sheet onto a tripod (a small metal frying pan could be used).

Place a beaker of water on top of the metal sheet.

- Record the starting temperature of the water.

Heat the base of the sheet with a Bunsen burner for one minute and measure the temperature.

- Work out the increase in temperature of the water.

Repeat the experiment with a heat resistant mat instead of the metal plate or pan.

- Which material was the best conductor?
- What did you do to ensure that this was a fair test?

 The equipment will become very hot. Allow it to cool properly before touching it

Expert tips

Remember that thermal energy moves from hot parts of a material to the colder parts.

 IGCSE Link...
You will learn more about the role of electrons in conduction in IGCSE Physics.

Explain the choice of the different materials used in the manufacture of a pan

Key terms

- **conduction**

Summary questions

1 Explain why an iced drink contained in a polystyrene cup will stay colder than the same drink in a metal cup.

2 Describe how thermal energy passes through a solid brick wall.

3 Design an experiment to find out which metal is the best conductor of thermal energy. You must make sure that the experiment is a fair test.

Gases and liquids are fluids. The particles they are composed of are free to move around. This means that thermal energy can be transferred from one place to another by the moving particles.

Expansion, contraction and density

When a material is heated, the particles inside the material gain energy, move more quickly and spread apart more. The particles occupy more space and the material expands.

Changes in density

When a block of metal is heated it will expand and so its volume will increase. However, the mass of the block will remain the same. As the density of the block equals its mass divided by volume, the density of the block must decrease as it gets hotter.

- Materials **expand** and become less dense when they are heated.
- Materials **contract** and become denser when they cool.

Floating and sinking

In a fluid the particles are free to move. This means that the less dense part of the fluid will float to the top leaving the denser (cooler) fluid at the bottom. This movement of particles will cause thermal energy to spread throughout the fluid in a **convection current**.

The fluid is heated > The fluid expands > The fluid becomes less dense > The fluid floats

Draw a flow chart similar to the one above describing what happens when a fluid cools.

Learning outcomes

After this topic you should be able to:

- describe the process of thermal energy transfer by convection in fluids
- explain how convection currents can be used to keep a room cool.

Expert tips

Remember that when a material expands, the particles do not get larger. The distance between the particles increases.

Exploring convection currents

When different parts of a fluid are at different temperatures, a convection current forms and transfers thermal energy through the fluid.

Using convection currents

In hot countries air conditioner units are placed at the top of a room to produce convection currents.

- The air conditioner cools some air at the top of the room.
- This cool air is denser than the rest of the air in the room and so sinks towards the floor.
- The cool air gradually warms up and expands, rising to the top of the room.
- As the air rises new, colder air produced by the air conditioner sinks downwards and replaces the warm air.

This process means that there will be a constant flow of cool air in the room.

An air conditioning unit is used to cool the air in a hotel room. The cool air will fall directly onto the bed on the left

 IGCSE Link...
You will learn more about the behaviour of particles in solids, liquids and gases in IGCSE Physics when you describe the changes in the forces between particles during changes of state.

Key terms

- **expand**
- **contract**
- **convection current**

Summary questions

1 Explain how convection transfers thermal energy in a fluid.

2 Draw a flow chart showing how convection currents form in a liquid as it is heated from the bottom.

3 A heater is switched on at ground level in a large warehouse. Why does it take a long time for the lower part of the room to become warm?

Infra-red radiation

Infra-red radiation is part of the electromagnetic spectrum. Infra-red radiation behaves in a very similar way to visible light. It can be emitted, absorbed or reflected.

Absorbing and emitting

Objects can heat up or cool down by the transfer of infra-red radiation:

- An object absorbing infra-red radiation will heat up as it gains energy.
- An object emitting infra-red radiation will cool down as it loses energy.
- When an object reflects infra-red radiation there is no change in temperature. The object has not lost or gained energy.

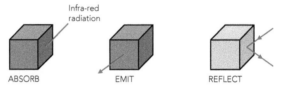

Absorbing, emitting or reflecting infra-red radiation

Cooling to reach a constant temperature

A hot object will emit infra-red radiation more rapidly than a cool object.

All objects emit and absorb some infra-red radiation. The hotter an object is, the more infra-red radiation it will emit to the surroundings. However, this hot object will also be absorbing a small amount of infra-red energy from the surroundings.

As the object cools, the amount of infra-red radiation it emits will decrease. Eventually the amount of infra-red radiation emitted by the object will be the same as the amount of energy absorbed by the object. The temperature of the object will stay the same. Objects at 'room temperature' are emitting and absorbing infra-red radiation at the same rate.

The effect of colours on heating and cooling

The colour of an object has a significant effect on the amount of infra-red energy it absorbs or emits.

- Dark colours are the best emitters and absorbers of infra-red radiation.
- Light colours are the worst emitters and absorbers of infra-red radiation.

Maintaining the temperature of buildings

The colour of buildings can affect how the buildings warm up during the day and cool down during the night. This is particularly important for regions where there is a large change in temperature from daytime to night-time such as deserts.

The white walls of these buildings will help keep them cooler in the day and warmer at night

During the day a lightly coloured building will slowly absorb infra-red energy. This means that the building will warm up slowly. During the night the building will slowly emit infra-red radiation. This means the building will slowly cool down.

Key terms

- **infra-red radiation**

Practical activity   Demonstrating the absorption of infra-red radiation

The effect of the surface colour on the absorption of infra-red radiation can be demonstrated.

Two metal surfaces are placed an equal distance from a radiant heater (electrical heating element).

One surface is matt black and the other surface is shiny silver.

A coin is attached to the back of each of the surfaces with wax.

- Predict what will happen to the temperature of the metal surfaces.
- Explain the result of the experiment in terms of absorption of infra-red radiation.

 The radiant heater will be very hot and the surfaces will become hot during the experiment.

Summary questions

1 Some frozen food is brought out of a freezer and allowed to reach room temperature. In what way does this process involve infra-red radiation and why does the temperature of the food stop rising when it reaches room temperature.

2 Nasrin wears dark clothing and goes out on a sunny day. When he stands in bright sunlight he feels very hot but when he stands in the shade he feels cold. Explain what is happening in these two situations.

3 Explain why houses in desert regions are not painted black. Give, and explain, two reasons.

When we get hot we sweat. Our bodies release water through pores and this water will evaporate, cooling the skin.

Explaining cooling by evaporation

Evaporation involves particles in a liquid escaping from the surface of the liquid and taking energy with them.

The particles in a liquid do not all have the same amount of energy. They have a range of different energies. The particles with the most energy will be moving fastest. If fast-moving particles reach the surface of the liquid, the particles can overcome the forces holding them in the liquid and can escape. The particles are now a gas or vapour.

When the fastest particles escape the average energy of the particles in the liquid will be reduced. Because the average energy has decreased, the temperature will also decrease.

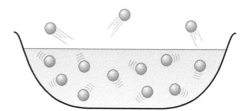

The most energetic particles can escape from the surface of a liquid causing the liquid to cool

Factors affecting cooling by evaporation

There are four factors that will affect the rate of evaporation:

1 Surface area – the greater the surface area of the liquid, the more particles will be near the surface. This makes it easier for particles to escape and evaporation will be faster.

2 Temperature of the liquid – the higher the temperature of a liquid, the greater the average energy of the particles will be. This means that the particles are more likely to escape and evaporation will be faster.

3 Air flow – when particles escape from a liquid they may stay in the region directly above the liquid and make it more difficult for more particles to escape. When air flows over the liquid the escaped particles are moved out of the way. This allows other particles to escape and speeds up evaporation.

4 Type of liquid – ethanol has a boiling point of 78 °C which is lower than the boiling point of water. This is because the attraction between ethanol molecules is weaker than the attraction between water molecules. Ethanol molecules can escape from the surface of the liquid more easily and so ethanol will evaporate more quickly than water at the same temperature.

The water in this salt pan will evaporate quickly as the surface area is large

Practical activity — Testing the factors that affect evaporation

Select a factor that affects the rate of evaporation to investigate.

Design an experiment that will show that your chosen factor has an effect on the rate of evaporation.

You must produce a detailed method and include how data will be recorded.

 Your plan must be checked by your teacher before you carry it out.

- What will you do to measure how much liquid has evaporated?
- What factors do you have to keep constant for the test to be fair?

 Do not use ethanol or any other flammable substance near a flame or electrical heating element.

Expert tips

Evaporation only happens at the surface of a liquid. Boiling happens throughout the liquid.

Key terms

- evaporation

Summary questions

1 A student spills a small amount of ethanol on a desk during an experiment and it spreads over the surface of the desk.
   a) Why does the ethanol on the desk evaporate quickly while the ethanol remaining in the bottle evaporates slowly?
   b) Some of the ethanol is spilt onto the student's hands. His hands feel cold when he blows on them. Why is this?

2 Why it is more difficult for an athlete to cool down on a humid day than a dry day?

Fuels contain chemical energy and are burned to provide us with useful energy in many ways. For example:

- in power stations the chemical energy in fuels is transformed into electricity
- in cars the chemical energy in the fuel is transformed into kinetic energy.

Burning fossil fuels

In many countries **fossil fuels** are the main source of energy. All fossil fuels contain large amounts of carbon. When the fuels are burned; thermal energy is released and large volumes of carbon dioxide gas are produced.

carbon + oxygen → carbon dioxide (+ ENERGY)

Formation of fossil fuels

Our fossil fuel sources of coal were formed about 300 million years ago. Coal was formed from plant material that did not decay and was preserved in swamps. Over millions of years the plant material was compressed by layers of rock being deposited on top. The pressure converted the plant material into layers of coal.

Oil and natural gas are also fossil fuels, formed around 150 million years ago. These fuels formed from tiny aquatic plants and animals that died and sank to the bottom of oceans. These were changed in a similar way to coal. Over millions of years oceans have shifted leaving many deposits of oil and natural gas beneath land.

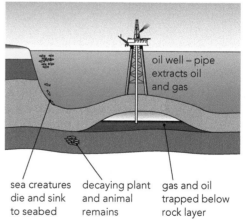

oil well – pipe extracts oil and gas

sea creatures die and sink to seabed

decaying plant and animal remains

gas and oil trapped below rock layer

The formation of oil and natural gas

Biofuels

Biological material (**biofuels**) can also be burned to release energy. Wood has been used for thousands of years for cooking and heating and some small power stations now use wood to produce electricity. Methane gas is produced when waste rots and this gas can be collected and burnt as a fuel.

Renewable and carbon-neutral fuels

Because biofuels are produced from plants they can be described as **carbon neutral**. The carbon released when the biofuels are burned is the same carbon that the plants absorbed from the atmosphere as they grew. No additional carbon is released. New fuel crops can be grown each year and so biofuels are **renewable**.

Biodiesel can be produced from plant crops and used in the same way as diesel from crude oil

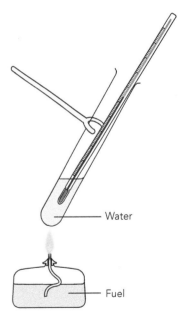

Water

Fuel

🔗 IGCSE Link...

You will learn more about measuring the efficiency of power stations when you study IGCSE Physics.

Key terms

- **fossil fuel**
- **biofuel**
- **carbon neutral**
- **renewable**

Practical activity	Measuring the energy provided by fuels

Measure the energy released by burning different liquid fuels.

Use a spirit burner to burn the fuel for a measured length of time.

Measure the change in temperature of a sample of water.

Weigh the burner before and after the fuel is burnt to measure out how much fuel has been used.

- Which fuel provides the most energy per gram?
- Compare your results with those of other students to find the reliability (reproducibility) of the experiment.

 Wear eye protection and be careful not to spill any fuel.

Summary questions

1 a) Which of the following fuels are renewable?

   beeswax, coal, wood, oil, petrol, paraffin wax

   b) Which gas is always produced when a carbon based fuel is burnt?

2 What advantages do biofuels have when compared with fossil fuels?

3 Fossil fuels are still being produced by dying plants and animals. Why aren't fossil fuels considered to be renewable?

IGCSE Link...

You may be able to investigate the power output of photovoltaic cells when you study IGSE Physics.

An engineer tests the output of a large array of photovoltaic cells

The Earth receives billions of joules of energy every second from the Sun. Scientists have developed a range of technologies that allow us to use this light energy.

Heating using sunlight

Solar furnaces

A **solar furnace** is designed to concentrate the energy from sunlight onto a small point. Small furnaces use concave mirrors to collect the sunlight that falls on a large area onto a small point. This energy can be used to cook or to boil water.

These curved mirrors are focusing the sunlight onto pipes containing water

Large solar furnaces have been built in deserts. Arrays of curved mirrors concentrate the sunlight onto a very small area and boil water to turn it into high pressure steam. This steam can be used to spin a turbine and generate electricity.

Solar heating

On some houses, black solar panels can be used to absorb the energy from sunlight. Water passes through these panels and absorbs energy. The hot water can be used in baths, showers or central heating systems. Solar panels reduce the need to use electricity or fossil fuels for heating.

Photovoltaic cells

Electricity can be generated directly from sunlight using **photovoltaic cells**. These contain semiconducting materials that produce a potential difference (p.d.) when exposed to light. This p.d. is used to produce a direct current that can be used to power electrical devices. The current can also be used to charge a battery to store the energy for later use.

Large photovoltaic panels are very expensive to manufacture. They are most efficient when used in locations where continuous sunlight will allow them to operate every day of the year.

Satellites in orbit around the Earth use photovoltaic cells to provide electrical energy as they have no place to store fuels.

Practical activity   Solar heating

Use a concave mirror to build a solar heater to investigate how effective it is. A concave mirror can be made from the inside of an umbrella or parasol lined with aluminium foil if you have no mirrors available.

Place a test tube containing water at the focal point of a concave mirror in bright sunlight.

Record the change in temperature of the water over a period of five minutes.

Compare the temperature rise to the temperature rise of a test tube of water placed in sunlight without the mirror.

- What colour should the test tube be in order to absorb most of the energy?

⚠ **Do not focus sunlight into your eyes, they can be permanently damaged.**

This solar cooker is being used to boil water on a mountaintop

Key terms

- **photovoltaic cell**
- **solar furnace**

Summary questions

1 Draw energy transfer diagrams for a solar furnace and a photovoltaic cell.

2 Some calculators have photovoltaic cells but these calculators usually have a battery as well. Why is this?

3 The following table shows the power output of a photovoltaic panel used to provide electrical energy to a house throughout the day.

Time	midnight	2 am	4 am	6 am	8 am	10 am	midday	2 pm	4 pm	6 pm	8 pm	10 pm
Power / W	0	0	0	50	120	150	170	150	130	100	30	0

a) Plot a graph comparing the power output to the time of day.

b) Why can't the panel be used to directly power the house lights during the night?

c) The panel is rectangular and measures 2.0 m by 1.5 m. What is the maximum power output per square meter of the photovoltaic panel?

The world trade centre building in Bahrain has several wind turbines built into its structure. The turbines produce up to 15% of the electricity requirements of the building

Wind power

Wind turbines collect the kinetic energy of wind from the air and transform it into electrical energy. The wide blades are pushed around by the wind. The blades are connected to a generator. As the blades spin the generator is turned and a potential difference is produced. This p.d. produces a current.

Wind turbines need to operate in a range of different wind speeds. They need complex gearing systems and electronics to produce a steady output frequency and p.d.

Wind turbines are fairly expensive to build but they do not require any fuel and so have low running costs. They can be quite noisy and so are usually built far from highly populated areas. Some wind turbines are built off-shore. These off-shore turbines are more expensive to install but are exposed to higher wind speeds.

Electricity from water
Hydroelectricity

Hydroelectricity uses the movement of water to generate electricity. Rivers are naturally supplied with water through the water cycle. Larger rivers can be dammed so that an artificial lake is produced. The water that builds up behind the dam is a store of gravitational potential energy. Water is released through pipes forcing it to pass through turbines. The turbines spin and generate electricity.

Large dams can produce vast amounts of electricity as millions of tonnes of water pass through them each hour. Small rivers can be used to generate electricity on a smaller scale.

This hydroelectric dam has a power output of millions of watts. Some dams are large enough to provide enough electricity for a large city

Building dams and the water reservoirs behind them requires a great deal of land and resources. The local habitats are destroyed and valuable farmland may be flooded.

Tidal power

Tidal power is another form of hydroelectricity. River estuaries fill up twice per day as the tide rises. If the estuary is dammed the water can be forced to flow through turbines and generate electricity. This tidal flow is a very reliable source of energy. However, the habitat in the river is dramatically changed when the tidal dam is built.

Wave power

The motion of waves can be used to drive a variety of different electrical generators. Although individual generators produce only small amounts of energy, thousands can be constructed around a coastline. Wave generators are expensive to install. They can also be difficult to maintain but they can provide a large source of renewable energy.

Expert tips

Nearly all electricity production relies on a turbine and generator combination. Try to spot these in all of the different designs.

Practical activity   Transferring energy to water

Construct a wind turbine from the equipment provided. Test the turbine to find out how many revolutions per minute it will complete in different wind speeds.

You can use a desk fan or hairdryer to produce the wind.

If you have a model wind turbine with a generator, you can measure the electrical output for different wind speeds.

- Do large wind turbines spin faster or slower than small ones?

Key terms

- **wind turbine**
- **hydroelectricity**

Summary questions

1 A team of engineers tested a range of wind turbine designs in a wind tunnel.

Wind speed:	1 m/s	4 m/s	8 m/s	12 m/s
Power output Model A	14 W	20 W	34 W	40 W
Power output Model B	12 W	18 W	34 W	50 W
Power output Model C	10 W	15 W	27 W	41 W

   a) What is the relationship between wind speed and the power output of the turbines?
   b) Which design operates best a low wind speeds?
   c) Which design operates best a high wind speeds?

2 Where does the energy that powers the water cycle originate?

3 The gravitational potential energy in 1 m³ of water behind a dam is 100 000 J. The water passes through a turbine at a rate of 600 m³ per minute.
   a) What is the maximum power output of the turbine?
   b) What is the maximum power output if the turbine is only 30% efficient?

Geothermal energy and nuclear fission

Geothermal energy

The interior of the Earth is very hot. In places on the Earth where the crust is at its thinnest, very high temperature rock can be found just a few kilometres beneath the surface.

Geothermal power plants pump water down to these hot rocks and the water is converted into high pressure steam. The steam returns to the surface and can be used to turn turbines and produce electricity.

Geothermal plants do not produce pollution but there are not many places on the Earth's surface where they can be built. Places near the plate boundaries are often the best locations but these may not be in populated areas.

A geothermal power station in Iceland. Geothermal energy provides Iceland with over 60% of its total energy needs including electricity and heating

The source of the Earth's thermal energy

The thermal energy within the Earth is a result of the breaking apart of the nuclei of atoms. The nuclei in the atoms of heavy elements such as uranium break apart and release thermal energy during radioactive decay. This is a natural process that releases vast amounts of energy over very long periods of time. Because these processes will continue for billions of years, geothermal energy is considered to be a renewable resource.

Nuclear fission power

Nuclear fission is the process used in all current nuclear power stations. Large nuclei are 'induced' to break apart and release energy inside a reactor core. The breakdown of one nucleus will give out neutrons that will induce other nuclei to break apart in a controlled chain reaction. The heat energy released by the nuclear reactions can be used to generate steam for turbines.

Nuclear waste and nuclear accidents

The products of nuclear decay in reactors are highly radioactive. Radioactive materials will damage living tissue, killing cells or causing mutations that can lead to cancers. The radioactive materials produced in a nuclear reactor will remain very dangerous for thousands of years. So the waste has to be stored carefully and securely.

Used nuclear fuel is highly radioactive and has to be kept cool

If a nuclear reactor overheats it can cause a 'meltdown' in which the reactor core ruptures. Then radioactive elements can escape into the atmosphere, contaminating a very large area. Only a few major nuclear accidents have occurred but the consequences have been severe.

Construction and decommissioning (taking apart) of nuclear power stations is very expensive. So the energy produced by this method is also costly.

An advantage of nuclear power is that no carbon dioxide (a greenhouse gas) is produced in the process. So generating electricity using nuclear fuel does not contribute to global warming.

> **IGCSE Link...**
> You will learn a lot more about radioactive decay in your study of IGCSE Physics. This includes the changes in the nucleus and nuclear decay equations.

Key terms

- **geothermal power**
- **nuclear fission**

Summary questions

1 What causes the high temperatures inside the Earth?

2 In what way is geothermal energy similar to nuclear fission power?

3 a) Why is electricity generated from nuclear power plants expensive?

  b) State an advantage of nuclear power compared with electricity generated by burning fossil fuels.

The demand for electricity and fuels continues to grow as the Earth's population increases and more societies become dependent on technology.

New power station technology

Carbon capture and storage

When a fossil fuel is burned, the carbon dioxide enters the atmosphere and contributes to global warming. Removing the carbon dioxide from waste gases after combustion is possible. This would reduce the impact of burning fossil fuels.

Carbon capture technology has been developed that would allow power stations to reduce the pollution they produce. The technology is expensive to develop and use but would allow the continued burning of cheap fossil fuels.

The captured carbon dioxide would need to be stored so that it could not escape into the atmosphere. It may be possible to pump the carbon dioxide into empty deep oil wells for long term storage.

Nuclear fusion

Nuclear fusion is the process that releases energy in stars. Light elements such as hydrogen or helium are 'fused' together producing heavier elements and releasing energy. Experimental power stations have been built to carry out nuclear fusion on the Earth. If fully operational nuclear fusion power stations can be built successfully, then an almost limitless source of electrical energy would be available.

In this fusion chamber hydrogen fuel is heated to the same temperature as the core of the Sun

IGCSE Link...

You will learn more of the details of nuclear fusion and fission when you study IGCSE Physics.

Nuclear fusion is a very difficult process to perform. A star can force the nuclei together because of the high pressure and temperature inside the core of the star. These conditions are difficult to produce on the Earth. So at present any nuclear fusion technology is still experimental.

Hydrogen fuel for vehicles

Scientists have been attempting to develop new fuels that do not contain carbon and so do not produce damaging carbon dioxide.

Hydrogen is the most common element in the universe and the Earth's oceans store a vast amount of hydrogen in water molecules (H_2O). When hydrogen is burnt in air the only product of the reaction is water vapour:

$$\text{hydrogen + oxygen} \rightarrow \text{water}$$

Practical activity Hydrogen combustion

Your teacher may demonstrate the combustion of hydrogen in oxygen on a small scale.

- What evidence was there that energy was released by the reaction?
- What evidence was there that water was produced?

 Never react large volumes of hydrogen and oxygen gases. The loud explosions can damage your ears.

Hydrogen fuel can store a large amount of chemical energy in a small volume. This high energy concentration means that the fuel can be used to power a wide range of vehicles from cars to planes. Hydrogen fuel makes an excellent replacement for petroleum based fuel.

Liquid hydrogen is used as a rocket fuel. The reaction with liquid oxygen releases vast amounts of energy and clouds of water vapour

Hydrogen needs to be extracted from water by electrolysis. Electrolysis requires large amounts of electrical energy and so producing hydrogen in this way does not save energy. More power stations would need to be built to produce the fuel in the first place.

Key terms

- **carbon capture**
- **nuclear fusion**

Summary questions

1 What advantages does hydrogen fuel have when compared with petroleum? What disadvantages does it have?

2 Why has it not been possible to make a fully functioning fusion reactor on the Earth?

3 Some scientists have suggested dissolving waste carbon dioxide in the oceans. What benefits and problems could this cause?

1 Which of these energy sources are renewable? [3]

    coal, nuclear fuels, wind, geothermal, oil, tidal

2 A block of ice is placed on a metal tray. The ice slowly melts.

  a Describe the processes by which the ice gains energy from the surroundings. [2]

  b Why does the metal tray become cold? [1]

3 A group of students is discussing the disadvantages of various energy sources. Which different energy source is being discussed when each of these points is raised?

  a Radioactive waste is produced. [1]

  b Large amounts of land would be flooded. [1]

  c They are very noisy. [1]

  d It will only work where the Earth's crust is thin. [1]

  e It needs large amounts of electricity to produce and so doesn't save energy. [1]

  f It produces large amounts of carbon dioxide. [1]

4 A group of students are studying the effect of surface colour on the cooling of containers. They monitor the temperature of two cups, one black and the other white, over a period of time.

  a Plot a graph of their results shown below. [3]

  b Make a conclusion based on the pattern in the results. [1]

  c Describe why the cups are cooling at different rates. [2]

5 A student has produced some copper sulfate solution and wishes to produce copper sulfate crystals. Describe what the students can do to increase the rate of evaporation of the solution without heating it? [3]

6 Some students are investigating the rate of evaporation of different liquids; water, ethanol and ether. The experiment is carried out in a fume cupboard. The students pour the liquids into identical rectangular trays measuring 4 cm (width) by 5 cm (length).

  a What is the surface area of the liquids in the trays? [1]

The students measure the mass of the liquids and trays at the start of the experiment and after ten minutes. The results are shown below.

Liquid	Starting mass / g	End mass / g	Change in mass / g
Water	14.4	14.2	
Ethanol	13.4	12.4	
Ether	15.8	12.6	

  b Which liquid has evaporated at the fastest rate? [1]

  c What is the rate of evaporation per cm^2 for each of the liquids? [3]

At the end of the experiment the students notice that they did not turn the fume cupboard extractor fan on as they were supposed to.

  d In what way would the results be different if the fan had been turned on? [1]

Time / s	0	30	60	90	120	150	180	210	240	270	300
Black cup temperature / °C	80	77	75	73	71	69	68	66	65	64	62
White cup temperature / °C	80	78	76	75	73	72	709	69	68	67	65

7   **a** Complete the following combustion
equations:

carbon + oxygen → _____   [1]

_____ + _____ → water   [2]

b What advantages does hydrogen fuel
have when compared with carbon based
fuels?   [2]

8 The world's total installed energy production
from photovoltaic cells is shown in this graph.

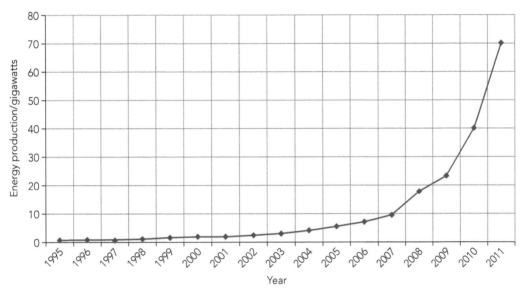

REN global renewables energy report 2012

a Describe the pattern in the graph.   [2]

b Estimate the energy production from
solar power in 2012.   [1]

Germany produces 35% of the world's solar
generated electricity.

c What was Germany's solar energy output
in 2011?   [1]

d What factors limit Germany's ability to
produce more solar electricity?   [2]

Glossary

A

Alternating current Current in which charge moves in alternating directions. Mains electricity provides alternating current.

Ammeter A meter used to measure the size of a current in amperes (A).

Ampere The unit of current. Symbol (A).

B

Battery A collection of cells working together to produce a larger potential difference.

Biofuel A fuel produced from plant or animal materials.

C

Carbon capture The capture of carbon dioxide produced in power stations so that the carbon does not enter the atmosphere.

Carbon neutral A fuel that does not add extra carbon dioxide to the atmosphere

Cell The chemical reactions in a cell provide the potential difference, which creates a current in a circuit.

Circuit diagram A diagram indicating how components are connected together.

Conduction The transfer of thermal energy from particle to particle through a material.

Contract A material becomes smaller and denser as the particles become closer together.

Convection current The flow of thermal energy through a fluid caused by the movement of particles from place to place.

Current A flow of charge. The rate of flow of charge is measured in amperes (A). The current in an electrical circuit is produced by a flow of electrons.

D

Digital sensor A sensor that produces digital signals that can be interpreted by computer systems.

Direct current Current in which charges only travel in one direction. Batteries produce direct current

E

Equilibrium When there is no resultant force or moment acting on an object, it is said to be in equilibrium.

Evaporation The escape of energetic particles from the surfaces of a liquid. Evaporation causes cooling of the remaining liquid.

Expand A material becomes larger and less dense as the particles become further apart.

F

Force multiplier A lever used to increase the size of the force.

Fossil fuel A fuel based on the fossilised remains of plants or animals.

G

Geothermal power Using the high temperatures within the Earth's crust to produce electricity.

H

Hydraulics machines Devices that use liquids to transmit forces.

Hydroelectricity The use of the gravitational potential energy in water to produce electricity.

I

Infra-red radiation A form of electromagnetic radiation similar to light but undetectable by the human eye. IR radiation can be reflected, absorbed and emitted.

M

Moment The turning effect caused by a force acting a distance from a pivot (moment = force × distance).

N

Negative charge The charge carried by electrons.

Nuclear fission The splitting of large atoms to release energy.

Nuclear fusion The release of energy when small atoms join together. This process occurs in the Sun.

P

Parallel circuit A circuit which has more than one route for the current to take.

Photovoltaic cell A device that produces electricity directly from sunlight.

Pivot The point about which a lever turns.

Positive charge The charge carried by protons.

Potential difference The 'push' that forces a current around a circuit.

Pressure A force acting over an area (pressure = force ÷ area).

R

Renewable An energy source that will not run out.

Resistance The resistance of a circuit reduces the size of the current in it.

S

Sensor A device which senses the environment in an electrical circuit.

Series circuit A circuit with only one route for the current to take.

Solar furnace A device that focuses sunlight on a spot to produce very high temperatures.

T

Temperature A measure of how hot or cold something is. Thermal energy will transfer from places at higher temperature to lower temperature. Temperatures are measured in °C (degrees Celsius).

Thermal energy The internal energy of an object due to the movement of the particles within it.

V

Volt The unit of potential difference. Symbol (V).

Voltmeter A meter used to measure the size of a potential difference in volts (V).

W

Wind turbine A device that converts the kinetic energy in wind into electrical energy.

Work An energy transfer caused by a force (work done = force × distance moved in the direction of the force).